Nicholas Rescher
Philosophy Examined

Nicholas Rescher
Philosophy Examined

Metaphilosophy in Pragmatic Perspective

DE GRUYTER

Contact Information:

Nicholas Rescher
Department of Philosophy
University of Pittsburgh
1032 Cathedral of Learning
Pittsburgh PA 15260

Tel: 412-624-5950
email: rescher@pitt.edu

ISBN 978-3-11-127625-0
e-ISBN (PDF) 978-3-11-074738-6
e-ISBN (EPUB) 978-3-11-074745-4

Library of Congress Control Number: 2021943960

Bibliographic information published by the Deutsche Nationalbibliothek
The Deutsche Nationalbibliothek lists this publication in the Deutsche Nationalbibliografie;
detailed bibliographic data are available on the Internet at http://dnb.dnb.de.

© 2023 Walter de Gruyter GmbH, Berlin/Boston
This volume is text- and page-identical with the hardback published in 2021.
Cover image: Altayb / iStock / Getty Images Plus
Printing and binding: CPI books GmbH, Leck

www.degruyter.com

For John Crosby

Contents

Introduction: The Mission of Philosophy? —— 1

1 Speculative Thinking —— 12

2 On Problems of Getting It Right —— 16

3 Contextual Acceptance and Cognitive Compartmentalization —— 29

4 Philosophical Methodology —— 33

5 Aporetic Method in Philosophy —— 47

6 Metaphilosophical Coherentism —— 64

7 On Philosophical Systematization —— 83

8 On Validating First Principles —— 110

9 Our View of Reality —— 133

10 Optimalism and Explanatory Totalization —— 137

11 Thematic Stability Amidst Philosophical Development —— 148

12 Philosophical Disagreement and Orientational Pluralism —— 156

13 Philosophical Cogency —— 187

14 The Pragmatic Perspective —— 198

Bibliography —— 211

Index of Names —— 215

Introduction: The Mission of Philosophy?

Philosophizing

Aristotle said it best. "Man by nature desires to know." The absence of information can be as distressing as that of food.

Philosophizing is a purposive enterprise that calls for addressing the "big question" of the human condition: man's place in the universe and the proper management of the obligations and opportunities of human life. It is a venture in rational inquiry that begins with problems and seeks solutions.

It is thus a purposive enterprise whose goal is to remove unknowing and puzzlement regarding the world and our place in its scheme of things. Answering "the big questions" is the name of the game. It is a project in answering pivotal questions about realty and its world on the basis of experience.

But we cannot contrive satisfactory answers out of thin air: to obtain credible conclusions we require premises. These will be the fruits f experience and can range from individual judgements and personal beliefs to universally valid information secured by impersonal "scientific" inquiry. But whose experience?

It all depends on the prevailing conditions of cognitive need. If we are content with answers that serve our personal requirements and suffice for self-satisfaction we can set the bar low. To satisfy our immediate colleagueship we must raise it. To meet the requirements of the community at large, we must raise it yet further.

Experience comes to us at three levels, the *person*, the *communal* (in the context of such key human prospects as the religious, the political, and the social, and the universal (i.e., science). Accordingly, there are three sorts of philosophers and philosophizers, depending on what level is prioritized. For these are the *philosopher's of opinion*. Their approach is "Here's how I see it, take it or leave it." (This includes thinkers such as Nietzsche or Heidegger.) Then there are the *Philosophers of culture*, whose approach is "Here is how we cultivated and right-minded sophisticates see it, surely you should do likewise." (This includes thinkers such as St. Thomas, Mill, Dewey, Cassirer.) Then there are the universalistic philosophers. Their approach is "This is how reality demands to be viewed by any rational person, anywhere, anytime." (This includes such thinkers as Plato, Aristotle, Kant.)

Note: This Introduction chapter is a slightly revised version of an essay that was originally published in Jacquette 2001.

Philosophy is identified as one particular human enterprise among others by its characterizing mission of providing satisfactory answers to the "big questions" that we have regarding the world's scheme of things and our place within it. And these big issues relate to fundamental of human concern, being universal in dealing with humans at large rather than particular groups thereof (farmers or doctors or Europeans or contemporaries of Shakespeare). Philosophical deliberations must have a bearing—direct or oblique—for the key essentials of the human condition—knowledge and truth, justice and morality, beauty and goodness, and the like.

In its dealing with such issues, philosophy principally asks questions having two forms:
- *Clarifactory* questions issuing from the format "Elucidate the nature of X (e.g., of truth, knowledge, justice)";
- *Explanatory* questions issuing from the format "Explain why P is so (e.g., why knowledge is not simply a matter of true belief)."

Either way, grappling with those "big questions" seeks to facilitate our understanding of the nature of things.

A philosophical discussion accordingly acquires importance by:
1. Its contribution to resolving a philosophical question that is itself (by the following standards) important:
2. Its resonance in other discussions that relate to it by way of support or refutation: its influence is the overall manifold of discussion;
3. Its impact on the question-agenda by raising new (but otherwise relevant) issues or by redirecting the stream of concern;
4. Its bearing in other (non-philosophical) issues by inspiring or influencing work in other domains of inquiry;

In sum, the importance of a philosophical discussion is manifested by its influence on the overall manifold of inquiry priority within philosophy itself.

The instruments of philosophizing are ideational resources of concepts and theories, and it deploys them in a quest for understanding, in the endeavor to create an edifice of thought able to provide us with an intellectual home that affords a habitable thought shelter in a complex and challenging world. The history of philosophy accordingly involves an ongoing intellectual struggle to develop ideas that render comprehensible the seemingly endless diversity and complexity that surrounds us on all sides.

As a venture in rational inquiry, philosophy seeks for the best available, the "rationally optimal," answers to our information-in-hand-transcending questions about how matters stand in the world. And experience-based conjecture

—theorizing if you will—is the most promising available instrument for question-resolution in the face of imperfect information. It is a tool for use by finite intelligences, providing them not with the best *possible* answer (in some rarified sense of this term), but with the best *available* answer, the putative best that one can manage to secure in the actually existing conditions in which we do and must conduct our epistemic labors.

In philosophy, as elsewhere throughout the domain of estimation, one confronts an inevitable risk of error. This risk takes two forms. On the one hand, we face errors of commission in possibly accepting what is false. On the other hand, we face errors of omission by failing to accept what is true. Like any other cognitive enterprise, philosophy has to navigate the difficult passage between ignorance and mistakes.

Two equally unacceptable extremes offer themselves at this stage. That first is to accept nothing, to fall into pervasive scepticism. Here we achieve a total exemption from errors of commission—but unfortunately do so at the expense of endless errors of omission. The other extreme is to fall into pervasive gullibility, to accept pretty much everything that is put before us. Here we achieve a total exemption from errors of omission—but unfortunately do so at the expense of maximal errors of commission. In philosophy, as in other branches of rational inquiry, we must strive for the best available middle way—the best available balance. Though we realize that there are no guarantees, we do desire and require reasonable estimates.

The need for such an estimative approach is easy to see. After all, we humans live in a world not of our making where we have to do the best we can with the limited means at our disposal. We must recognize that there is no prospect of assessing the truth—or presumptive truth—of claims (be they philosophical or scientific) independently of the use of our imperfect mechanisms of inquiry and systematization. And here it is *estimation* that affords the best means for doing the job. We are not—and presumably will never be—in a position to stake totally secure claims to the definitive truth regarding those great issues of philosophical interest. But we certainly can—and indeed must—do the best we can to achieve a reasonable *estimate* of the truth. We can and do *aim* at the truth in our inquiries, even in circumstances where we cannot make failproof pretensions to its attainment, and where we have no alternative but to settle for the *best available estimate* of the truth of the matter—that estimate for which the best case can be made out according to the appropriate standards of rational cogency.

The Systematization of Data

Yet despite those guarding qualifications about feasibility and practicability, the "best available" answer at issue here is intended in a rather strong sense. We want not just an "answer" of some sort, but a viable and acceptable answer—one to whose tenability we are willing to commit ourselves. The rational conjecture at issue is not to be a matter of *mere guesswork,* but one of *responsible estimation* in a strict sense of the term. It is not *just* an estimate of the true answer that we want, but an estimate that is sensible and defensible: *tenable,* in short. We may need to resort to more information than is actually given, but we do not want to make it up "out of thin air." The provision of reasonable warrant for rational assurance is the object of the enterprise. Rational inquiry is a matter of doing no more—but also no less—than the best we can manage to realize in its prevailing epistemic circumstances. Nevertheless, the fact remains that the rationally indicated answer does in fact afford our most promising *estimate* of the true answer—that for whose acceptance as true the optimal overall case be constructed in the circumstances at hand.

Now with regard to those "big issues" that constitute the agenda of philosophy the systematization in point of grounding and connectivity of otherwise available information is the best policy. Such *systematization* in the context of the available background information is nothing other than the process for making out this rationally best case. It is thus rational conjecture as based on and emerging from systematic considerations that is the key method of philosophical inquiry, affording our best hope for obtaining cogent answers to the questions that confront us in this domain. Let us consider more closely just what is involved here.

In philosophizing we strive for rational coherence in achieving answers to our questions. But how is one to proceed in this venture? It is clear that here, as in other branches of inquiry, we begin with data.

Neither individually nor collectively do we humans begin our cognitive quest empty handed, equipped with only a blank tablet. Be it as single individuals or as entire generations, we always begin with a diversified cognitive heritage, falling heir to that great mass of information and misinformation that the knowledge of our predecessors—or those among them to whom we choose to listen. What William James called our "funded experience" of the world's ways—of its nature and our place within it—constitute the *data* at philosophy's disposal in its endeavor to accomplish its question-resolving work. These specifically include:

- Common-sense beliefs, common knowledge, and what have been "the ordinary convictions of the plain man" since time immemorial;

- The facts (or purported facts) afforded by the science of the day; the views of well-informed "experts" and "authorities";
- The lessons we derive from our dealings with the world in everyday life;
- The received opinions that constitute the worldview of the day; views that accord with the "spirit of the times" and the ambient convictions of one's cultural context;
- Tradition, inherited lore, and ancestral wisdom (including religious tradition);
- The "teachings of history" as best we can discern them.

There is no clear limit to the scope of philosophy's potentially useful data. The lessons of human experience in all of its cognitive dimensions afford the materials of philosophy. No plausible source of information about how matters stand in the world fails to bring grist to the mill. The whole range of the (purportedly) established "facts of experience" furnishes the extra-philosophical inputs for our philosophizing—the potentially usable materials, as it were, for our philosophical reflections.

And all of these data have much to be said for them: common sense, tradition, general belief, and plausible prior theorizing—the sum total of the different sectors of "our experience." They all merit consideration: all exert some degree of cognitive pressure in having a claim upon us. Yet while those data deserve respect they do not deserve acceptance. And they certainly do not constitute established knowledge. There is nothing sacred and sacrosanct about them. For, taken as a whole, the data are too much for tenability—collectively they generally run into conflicts and contradictions. The long and short of it is that the data of philosophy constitute a plethora of fact (or purported fact) so ample as to threaten to sink any ship that carries so heavy a cargo. The constraint they put upon us is thus not peremptory and absolute—they do not represent certainties to which we must cling at all costs. Even the plainest of "plain facts" can be questioned, as indeed some of them must be, since in the aggregate they are collectively inconsistent.

And this is the condition of philosophy's data in general. We confront those data spreading out all around us by way of belief inclinations. But—as already stressed—they are by no means unproblematic. The constraint they put upon us is not peremptory and absolute—they do not represent certainties to which we must cling at all costs. For the philosopher, nothing is absolutely sacred. The difficulty is—and always has been—that the data of philosophy afford an embarrassment of riches. They engender a situation of cognitive over-commitment within which inconsistencies arise. For they are not only manifold and diversified but invariably yield discordant results. Taken altogether in their grand totality, the data are inconsistent. And here philosophy finds its work cut out for it.

In philosophy, we cannot accept all those "givens" as certified facts that must be endorsed wholly and unqualifiedly. Every datum is defeasible—anything might in the final analysis have to be abandoned, whatever its source: science, common sense, common knowledge, the while lot. Those data are not truths but only plausibilities. Nothing about them is immune to criticism and possible rejection; everything is potentially at risk. One recent theorist writes,

> No philosophical, or any other, theory can provide a view which violates common sense and remain logically consistent. For the truth of common sense is assumed by all theories. ... This necessity to conform to common sense establishes a constraint upon the interpretations philosophical theories can offer.[1]

But this is very problematic. The landscape of philosophical history is littered with theories that tread common sense underfoot. There are no sacred cows in philosophy—common sense least of all. As philosophy goes about its work of rendering our beliefs coherent, something to which we are deeply attached will have to give, and we can never say at the outset where the blow will or will not fall. Systemic considerations may in the end lead to difficulties at any point.

For these data do indeed all have some degree of merit and, given our cognitive situation, it would be very convenient if they turned out to be true. Philosophy cannot simply turn his back on these data without further ado. Its methodology must be one of damage control and salvage. For as regards those data, it should always be our goal to save as much as we coherently can.

Metaphilosophical Issues

To this point, the tenor of the discussion has been to offer a series of assertions along the lines of: This is what philosophy is; this is what philosophy does; this is how philosophizing works. But what justifies this way of talking? What reason is there to think that matters indeed stand as claimed?

This is a question that can, in the final analysis, be answered only *genetically*, by linking the response to and duly coordinating it with the historical facts about how philosophizing has actually been carried on over the years. What philosophy is all about is not writ large in the lineaments of theory but is something

---

1 Kekes 1980, p. 196.

that must be gleaned from the inspectable realities of philosophical practice. And so, while the history of physics may be largely irrelevant for physicists, the history of philosophy is unavoidably relevant for philosophers. What philosophers *should* do has to emerge from a critical analysis of what philosophers *have been* doing. The history of philosophy is not a part of philosophy, but philosophy cannot get on without it.

All the same, it is lamentable that now, more than 200 years later, there are still philosophers whose modus operandi invites Kant's classic complaint (at the start of the Introduction of the *Prolegomena*) that "there are scholarly men for whom the history of philosophy (both ancient and modern) is philosophy itself." For the fact is that philosophy and history-of-philosophy *address different questions*—in the one former instance what is the case about an issue, and in the latter what someone, X, thought to be the case. To address the former question we must speak on our own account. A philosopher cannot be a commission agent trading in the doctrines of others; in the final analysis he must deal on his own account. There must be a shift from "X thinks that A is the answer to the question Q" to the position that we ourselves are prepared to endorse for substantively cogent reasons. No amount of exposition and clarification regarding the thought of X and of Y will themselves answer the question on our agenda. To do so we must decide not what people thought or meant but what is correct with respect to the issues. And so while the history of philosophy is indeed an indispensable instrument of philosophy—in a science of concepts, ideas, problems, issues, theories, etc.—these are no more than *data* for our philosophizing. Actually to philosophize we must do more than note and consolidate such data, we must appraise and evaluate them on our own account. Philosophers must speak for themselves and conduct their business on their own account. They cannot hide themselves behind what X thinks or what Y thinks, but must in the end present a position of their own with respect to *what is to be thought*. The history of philosophy is not—and cannot be—a substitute for philosophy itself.

Metaphilosophy

Metaphilosophy—the study of the nature and methodology of the discipline—is also an integral component of philosophy. Unlike the situation with chemistry or with physiology, questions about the nature of *philosophy* belong to the discipline itself. And so, these questions about methodology cannot really be resolved by recourse to some sort of philosophy-neutral methodology. Only at the end of the day—only when we have pursued our philosophical inquiries to an adequate stage of development—will it become possible to see, with the wisdom of a more synoptic hindsight, as it were, that the selection of a methodological starting point was in fact proper and appropriate. It is part and parcel of the coherentist nature of philosophical method that our analysis must issue in smoothly self-supportive cycles and climates. Circularity in philosophical argumentation is not necessarily vicious. On the contrary, it can and should exhibit the ultimately self-supportive nature of rational inquiry at large. Herein lies a key part of the reason why philosophy must be developed systematically—i.e., as a system.

If you cannot fit your philosophical contentions into a smooth systemic unison with what you otherwise know then there is something seriously amiss with them. To be sure, this does not mean that the discussion will not here and there be projected into contentions that are controversial and seemingly eccentric. For sometimes the best reason for adopting a controversial and apparently strange thesis is that it contributes significantly to the systemic coordination of the familiar by serving to unify and rationalize a mass of material much of which seems comparatively unproblematic. For example, our basic thesis that philosophy exists to make sense of the things we know is far from being a philosophical truism. But that does not preclude its ultimate appropriations.

Metaphilosophy's key lesson is that the cardinal task of philosophy to impart systemic order into the domain of relevant data; to render them consistent, compatible, and smoothly coordinated. Its commitment to instilling harmonious coherence into the manifold of our putative knowledge means that systematization is the prime and principal instrument of philosophical methodology. One might, in fact, define philosophy as the rational systematization of our thoughts on basic issues—of the "basic principles" of our understanding of the world and our place within it. We become involved in philosophy in our endeavor to make systemic sense of the extraphilosophical "facts"—when we try to answer those big questions by systematizing what we think we know about the world, pushing our "knowledge" to its ultimate conclusions and combining items usually kept in convenient separation. Philosophy is the policeman of thought, as it were, the agent for maintaining law and order in our cognitive endeavors.

To this point, the tenor of the discussion has been to offer a series of assertions along the lines of: This is what philosophy is; this is what philosophy does; this is how philosophizing works. But what justifies this way of talking? What reason is there to think that matters indeed stand as claimed?

This is a question that can, in the final analysis, be answered only *genetically*, by linking the response to and duly coordinating it with the historical facts about how philosophizing has actually been carried on over the years. What philosophy is all about is not writ large in the lineaments of theory but is something that must be gleaned from the inspectable realities of philosophical practice.

And so, while the history of physics may be largely irrelevant for physicists, the history of philosophy is unavoidably relevant for philosophers. What philosophers *should* do has to emerge from a critical analysis of what philosophers *have been* doing. The history of philosophy is not a part of philosophy, but philosophy cannot get on without it.

All the same, it is lamentable that now, more than 200 years later, there are still philosophers whose modus operandi invites Kant's classic complaint (at the start of the Introduction has *Prolegomena to Any Future Metaphysic*) that "there are scholarly men for whom the history of philosophy (both ancient and modern) is philosophy itself." For the fact is that philosophy and history-of-philosophy *address different questions*—in the one former instance what is the case about an issue, and in the latter what someone, X, thought to be the case. To address the former question we must speak on our own account. A philosopher cannot be a commission agent trading in the doctrines of others; in the final analysis he must deal on his own account. There must be a shift from "X thinks that A is the answer to the question Q" to the position that we ourselves are prepared to endorse for substantively cogent reasons. No amount of exposition and clarification regarding the thought of X and Y will themselves answer the question on our agenda. To do so we must decide not what people thought or meant but what is correct with respect to the issues. And so while the history of philosophy is indeed an indispensable instrument of philosophy—in a science of concepts, ideas, problems, issues, theories, etc.—these are no more than *data* for our philosophizing. Actually to philosophize we must do more than note and consolidate such data, we must appraise and evaluate them on our own account. Philosophers must speak for themselves. They cannot hide themselves behind what X thinks or what Y thinks, but must in the end present a position of their own with respect to *what is to be thought*. The history of philosophy is not—and cannot be—a substitute for philosophy itself.

On the other hand, the fact is that metaphilosophy—the study of the nature and methodology of the discipline—is also an integral component of philosophy. Unlike the situation with chemistry or with physiology, questions about the nature of *philosophy* belong to the discipline itself. And so, these questions about methodology cannot really be resolved by recourse to some sort of philosophy-neutral methodology. Only at the end of the day—only when we have pursued our philosophical inquiries to an adequate stage of development—will it become possible to see, with the wisdom of a more synoptic hindsight, as it were, that the selection of a methodological starting point was in fact proper and appropriate. It is part and parcel of the coherentist nature of philosophical method that our analysis must issue in smoothly self-supportive cycles and climates. Circularity in philosophical argumentation is not necessarily vicious. On the contrary, it can and should exhibit the ultimately self-supportive nature of rational inquiry at large. Herein lies a key part of the reason why philosophy must be developed systematically—i.e., as a system.

If you cannot fit your philosophical claims into a smooth systemic unison with what you otherwise know then there is something seriously amiss with them. To be sure, this does not mean that the discussion will not here and there be projected into contentions that are controversial and seemingly eccentric. For sometimes the best reason for adopting a controversial and apparently strange thesis is that it contributes significantly to the systemic coordination of the familiar by serving to unify and rationalize a mass of material much of which seems comparatively unproblematic. For example, our basic thesis that philosophy exists to make sense of the things we know is far from being a philosophical truism. But that does not preclude its ultimate appropriations

What constitutes progress in philosophical inquiry? The issue is complicated. One is tempted to say: getting our questions answered. But this doesn't quite work. If anything the reverse is the case because in answering philosophical questions we almost always raise further ones. Exactly because philosophical issues are both convoluted and many sided, progress consists on unraveling the issues in ways that bring their complexity to light in processes of dialectical exfoliation. Philosophical progress consists in the ongoing removal of oversimplification—in conflicting the dialectical process of ongoing sophisticated approximation through which the never-ending complications of the issues is brought to light. The ongoing removal of errors of omission and commission in which the shortcoming of earlier effort are overcome.

The cardinal task of philosophy is thus to impart systemic order into the domain of relevant data; to render them consistent, compatible, and smoothly coordinated. Its commitment to instilling harmonious coherence into the manifold of our putative knowledge means that systematization is the prime and principal instrument of philosophical methodology. One might, in fact, define philosophy as the rational systematization of our thoughts on basic issues—of the "basic principles" of our understanding of the world and our place within it. We become involved in philosophy in our endeavor to make systemic sense of the extraphilosophical "facts"—when we try to answer those big questions by systematizing what we think we know about the world, pushing our "knowledge" to its ultimate conclusions and combining items usually kept in convenient separation. Philosophy is the policeman of thought, as it were, the agent for maintaining law and order in our cognitive endeavors.

This perspective provides the outlook of the present deliberations. It sees philosophizing as a goal-directed endeavor, and to heed a situation-coordinate requirement. But throughout the sphere of speculative inquiry—and in philosophy in particular—an admixture of practical concern cannot be avoided in securing the premises of theoretical deliberation.

To proceed rationally under the existing conditions philosophy must therefore take account of the pragmatic dimension of resource and requirement. Its proceedings are to be rational all right but in conditions where rationality is called on to house the conditions of practicality. And so in the final analysis philosophers can and must have a pragmatic, purposive, and goal oriented aspect—in short it must have an ineradicably pragmatic dimension.

How all this works itself out within the parameters of the descriptive is the object of concern in the deliberations of this book.

There are two variant modes of progress. When there is a definitive distinction we can measure progress either by moving from the start and by nearing the goal: The two come to the same thing. But where there is no definite distinction but only an aspirational goal we cannot determine distance from it but only respective distance away from the starting point. We can only measure progress by distance transversed. And this of course is the situation of philosophy. We can remove imperfections but not achieve perfection, can make improvements but not accomplish completion. It is this realization of work unaccomplished despite progress made that underlies the pragmatic aspect of realistic philosophizing. It is the guiding recognition that we take in the full acknowledgement that the idea is beyond us and that here, as elsewhere, we have to rest satisfied with the recognition that the best we can do has to be accounted as good enough.

1 Speculative Thinking

Speculation

"What if?" is the doorway to speculative thinking. For as human mentality matured over the ages, its range was decreasingly limited to factualities, with truths and falsities no longer the prime objects of its consideration. We came to be in a position to consider situations we neither accept nor reject but simply contemplate as instructive possibilities. From the cognitive point of view we became amphibious creatures, with minds that can address (be it correctly or not) not only the realm of the real but also that of mere supposition.

"What if" suppositions are matters of factual pretense, of cognitive play-acting: they state not actual facts but mere assumptions and hypothesis. Unlike assertions and postulations the do not affirm what is or is taken to be so, but only what is taken provisionally and pro tem—not for purposes of affirmation and information but only for purposes of consideration and exploration. The response to "what if?" questions differ from "since-because" claims in that they suspend belief in the antecedent. Here the critical thing is mere supposition, a proceeding that dispenses with belief.

The speculative wonderment of "What if?" paves the way to the imperatively conditional: "If-then." For the proper response to what-if questions is generally of the format if-then. (Question: "What if he misses the train?" Answer: "If he misses the train, he'll miss the boat.") And what makes such an answer correct is the linkage between its antecedent and its consequent, be it empirical and contingent as in the train/boat example or a matter of logico-acceptability necessary. (Question. "What if one of the five people present leaves?" Answer: "If one of the five people present leaves, then only four people will remain.") Thus in answering a "what if" question by a response of the format "If – – – then ...", we are looking to the consequences of that dash-indicated antecedent.

Consider for example the following sequence of what-if reasoning:

> The room is empty. [Fact]
> Suppose someone were in that room. [Supposition]
> If someone were in the room they would have had to climb in the window. [If-then consequence of supposition.]
> Why? Because the only door has been sealed shut. [Fact]
> But what if the window were also sealed shut? [Follow-up supposition]

It is clear that once we embark on speculation there is no end to the process. Just as with facts we can contrive unendingly to seek for the reason why behind the reason why, so with suppositions we can inadequately continue to wonder along the paths of possibility.

"What-if" thinking provides a vastly versatile instrument. In scientific contexts, it is a guide to experimentation ("What if we X and Y at a high temperature?"), in everyday contexts a goal to planning ("What if a misfortune X were to occur, how could we protect ourselves against its consequences); or it could be an instrument of criminal investigation ("If X committed a crime, what clues and indications might yet remain to show this?").

A key issue to address here is that of the semantical condition and status of the conditional antecedent. It must, of course, be *meaningful* otherwise we don't know what we are about. But it lies in the very nature of end hypotheticals that it need certainly not be *true*. Nor even need it even be something that is *possible*. There is virtually no limit to supposition. Beyond meaningfulness, assumptions are subject to no further restrictions. Not even logical self-consistency is required here, seeing that *ad absurdum* reasoning provides a clear counter-example.

As such, supposition is a resource of virtually unlimited scope. One can make suppositions about virtually anything—not just matters of conceivable fact, but also questions, injunctions, instructions, actions, problems. However, we shall here focus on fact-purporting suppositions on the order of: "Suppose that . . . were so, then - - -." And the affirmation at issue in the antecedent need not be an accepted truth. Its truth-status may be either false or unknown and problematic.

More on Prioritization

A crucial difference is operative when new data are introduced by way of discovery (finding and observation) rather than by way of supposition (hypothesis and assumption). For in the former case if the data don't fit it is they themselves that will need readjustment. But assumptions are sacrosanct. Other, preexisting materials have to be revised to make room for them. Reciprocal adjustments is the instrument of consistency in mandating either way. An optimal landscape of reciprocal accommodation has to be achieved one way or another via principles of rational economy.

Paradox analysis affords another illustration of this phenomenon. Thus suppose we adopt the underlying principle at issue in the classic "Paradox of the Heap":

> "If n sand-grains do not constitute a heap, neither do $n + 1$."

And in the wake of this supposition we are confronted with two further concepts:

> One or two sand-grains do not constitute a heap
> A million sand-grans constitute a big heap

It is clear that the sequential addition of sand-grains will engender a contradiction here, seeing that we realize full well that eventually a heap is before us. So we arrive at the puzzle: "What if to a single grain of sand we keep adding more. Just when do we arrive as a heap?"

As noted above, the priority order of claims on retention in the face of counterfactual supposition runs as follows: (1) the assumptions and suppositions whose postulations are at issue, (2) definitions and terminological explanations, (3) general laws, rules, principles, (4) speculative contingent factualities, (5) plausibilities and probabilities. Such a hierarch of epistemic-tenscity status obtains and serves to determine the weakest link at which to break the cycle of inconsistency that counterfactual suppositions engender.

In the wake of such considerations we have to reject that seemingly plausible addition principle in the face of more fundamental factualities. But of course the question "At just what point is the transition from non-heaps to heaps made?" is one that has no answer; it is inappropriate on grounds of resting as on mistaken presupposition. Transition there is and must be, but it is not punctiform. In this regard the question is based on the erroneous presuppositions of potential exactitude.

One useful and very common use of thought experimentation relates to its *explanatory* employment. We here reason along the lines of "If only such-and-such were the case, then something-or-other (which otherwise would be very difficult to explain) now admits of a ready and satisfying explanation." For example, Thales (6th century B.C.), the very first of the nature philosophers of ancient Greece, proposed to explain the annual flooding of the Nile as the result of a backing up of its outflow due to the opposing force of the annually recurrent Etesian winds.

Thought Experiments

Overall, a proper thought experiment involves five stages; supposition, context specification, commitment adjustment, conclusion deriving, and lesson drawing. And at each of these stages a mishap or malfunction can in theory arise.

- The supposition can turn out to be meaningless;
- The context may be set up inappropriately, in relation to the purposes of the thought experiment in particular by way of error of omission;
- The commitment adjustment may fail to be realistic, in particular by way of errors of omission that plunge matters into inconsistency;
- The course of reasoning by which the intended conclusion is drawn may be flawed and erroneous;
- The wrong lesson can be drawn for the experiment by overlooking possibilities for its interpretation.

In sum, all sorts of procedural flaws can in theory arise to vitiate a thought experiment.

Philosophical Speculation

Science addresses the real world. Granted it too engages in unreal and sometimes even counter-factual speculating, but always only to illustrate the ways of the real. Philosophy, by compassion, throws in speculation. It acknowledges that one can often best understand the nature and meaning of reality by considering the ways and means of what is not. Even as we often most acutely appreciate life's benefits when adverse circumstances compels us to do without them, so the contemplation of what is not can lead not to understand and appreciate the offerings and arrangements of the real. If a second man banished the doctrines and bathed our planet in eternal light, organic life or rather as we know it would become impossible. Since the very start of philosophy, thought experimentation has been one of its main instrumentalities.[2]

---

[2] For a more detailed elaboration, see Rescher 1991, pp. 31–41.

2 On Problems of Getting It Right

Getting It Wrong

It is instructive to approach the matter from the negative side, in considering the problems of getting it right in matters of belief and acceptance and begin with considerations about getting it wrong. Getting it wrong—error—is a matter of mistaken judgment, of accepting an incorrect answer to a question. It occurs principally in two domains, that of cognition and that of action, of wrong thinking and of wrong doing.

Wrong belief exacts its penalty by way of misinformation. But apart from such *cognitive* penalties for mistaken information there are also the *substantive* penalties for incorrect acting. And since homo sapiens as a rational being bases his actions on his beliefs, and acts with a view to putative consequences, this means that generally even accepting incorrect information is apt to exact a substantive penalty.

Even as the arrow can win the bull's eye by more or by less, so error of every kind can be more or less off the mark. And there are primarily two sorts of errors:
1. *Errors of the first kind:* Errors of commission arising when we mistakenly accept as true what is actually false. So what results is a wrong picture regarding the facts of the situation;
2. *Errors of the second kind:* Errors of omission arising when we inappropriately leave things out of manifold of acceptance. Here we mistakenly deem false what is actually true. So what results is a wrong picture as to the details of the situation.

Correspondingly, two sorts of error (error of information):
- Errors of truth (defects of correctness);
- Errors of informativeness (defects of accuracy and detail).

And the two modes of error (commission/likelihood and commission/detail) correlated with two prime aspects of information, namely
- *Likelihood:* strength of substantiation and evidentiation;
- *Detail:* precision of informativeness.

Evidential security is a relatively familiar epistemic issue; detail and precision rather less so. Both are critically important at issues in philosophy as elsewhere.

Assurance

Crucial for the theory of cognition in general—and for philosophical speculation in particular—is the concept of "attachment" to what is deemed to some extent acceptable. This comes in two forms:
- *Affective* attachment: this reflects the extent of like or dislike, our welcoming or unwelcoming of the (supposed) fact at issue;
- *Epistemic* attachment: this reflects the extent of assurances, conviction, or certainty that we attach to the fact at issue. A matter not of how probable we deem it to be, but of how difficult we would find it to abandon this belief.

The fact of these is irrelevant for present purposes. The second yield of the idea of an epistemic hierarchy of cognitive commitment. This reflects the extent to which we would find it problematic (difficult) to reject this contention, and yields a cognitive hierarchy of roughly the following order of epistemic precedence and priority:
- Definitions and meaning explanations;
- Lawful generalizations;
- Common knowledge;
- Reports regarding matters of contingent facts and particular observations.

This scale of presence comes into operation in all contexts of forced choice regarding what to retain and what to reject.

Epistemic assurance—the extent to which claims can be seen as meriting acceptance and endorsement—comes in a hierarchy of different *acceptability levels*, as we will call them. These levels of acceptability range somewhat as follows:
A1 With assured certainty;
A2 With reasonable assurance;
A3 With plausible grounding;
A4 With reasonable hope and expectation.

As one moves down the descending steps of such a hierarchy one encounters the phenomenon of *epistemic degradation* as the acceptability of the assertions declines and the risk of error increases.

A highly significant phenomenon comes into view at this point. The contentions we accept with assured certainty must all be mutually consistent as will any inferences that can be determined for them by logical proceedings. But where we leave this secure sign of $A1$—and increasingly as we move down that Ai ladder—there will be inconsistencies. (This conclusions that can reasonably or plausibly be drawn from certain premisses need not be reciprocally compatible and consistent.) And this means that different individuals who—for one reason or another—render the issue of consistency-resolution differently—can (in a perfectly reasonable way) resolve the issues differently and arrive at different (but equally justified)

answers to questions and positions on issues. In sum, increasing disagreement can arise in ways that are altogether natural and reasonable as one moves down the Ai ladder.

Detail

Getting it right by way of correctness (i.e., truth-likelihood) can always be embraced at the cost of detailed information.

We can always purchase assertoric safety—likely correctness—at the price of qualification and its corresponding loss in accuracy, detail, and informativeness. Fundamental in epistemology is a trade-off relationship of informativeness determination.

The detail/likelihood interrelation is depicted in Figure 2.1 characterizing. Informativeness combines the two: possible correctness and further detail. On most issues it is only available to a certain extent the $A = L \times D = c$ (constant) sets a limiting barrage. One of its dimensions can only be embraced at the expense of foregoing the other. A number of further epistemologically significant things happen as we step down such a ladder. The first and foremost is an exponential increase in the manifold of claimability. There are a great many more things one can claim and a great many more questions we answer if for usable premises we ask only for reasonable estimates rather than assured certainly, or again only for plausibility grounded evidentiation rather than firm assurance, and so on. The more we are willing to relax that standard of assurance of what we can say on a topic, the ampler and more fully rounded picture can we draw regarding the situation before us. The resultant situation may be characterizes as shown in Figure 2.1.

Epistemic Limits

With information as elsewhere there is a complementary between quantity and quality (see Figure 2.2). Any given question-issue (problem)—and so problem-area as well—has an *epistemic limit* set by a value of c in the EF (epistemic formula): $I \approx Q \times A = c$ (constant). The limiting value c is the *information threshold*

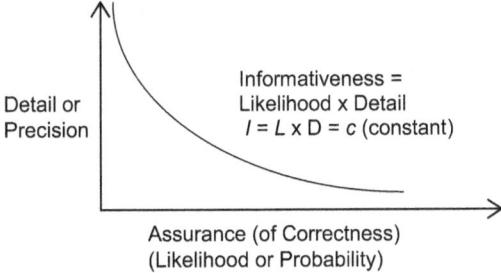

Figure 2.1: The Detail/Assurance Relation

for a given range of inquiry. It is a salient descriptive characteristic for the field at issue.

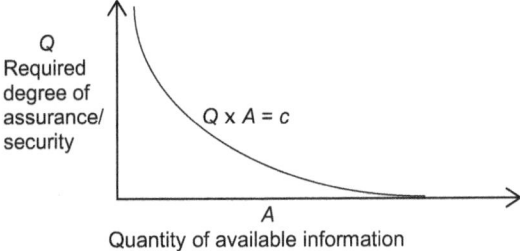

Figure 2.2: The Amount-Quality Relation

The c-number of a field of inquiry is an indicator of the extent to which its generalizations can dispense with qualifications heeded to provide for "exceptions to the rule"—the extent to which it can say how phenomena always and invariably rather than generally, usually, normal, and for the most part. In sum, it is an index of field's departure from descriptive and explanatory ease and regularity. For example, the c-value of mathematics is small, of economics comparatively large, and for philosophy pretty huge. (So if, say, mathematics is Z, economics is some 5Z, and philosophy some 30Z.)

This difference is reflected in the amount of compilation we need for characterizing the acceptability of luck the field's generalizations. For mathematics: one will do, namely categorically true. For economics we would need at least three: well-established, substantially confirmed and reasonably indicated. But significant generalization in philosophy could barely be characterized as outright truth or even well-established contentions but will have to settle for such qualifications as reasonable-suppositions, plausible conjectures, or widely-accepted contentions.

C-reduction as an ideal: but only realizable "up to a point."

The Issue of Cost: Paying For Quality

Consider an exceptionless and unqualified generalization such as

> All *A*s and *B*s

It is clear that this is more problematic—more demanding and more difficult to establish—than such analogous as:

> Almost all *A*s are *B*s
> Generally *A*s are *B*s
> *A*s are usually *B*s
> *A*s are normally *B*s

It is clear that all such qualifications can provide effective protection against error. It is, after all, far easier to go wrong with "all" than with "most," with "always" that with "generally" with.

Even if that initial generalization is erroneous and incorrect, its informative substance can in some measure be presumed by means of those qualifications.

Of course, the lesser in scope—simpler and more straightforward—those saving qualifications will be, the less the error of that initial claim.

Quality information costs. (Think back here to Figure 2.1's Detail/Likelihood relation: $I = L \times D = c$ (constant).)

Decreasing that constant c requires a significant improvement in the state-of-the-art of inquiry. And this not only has costs, but also limits. In any given area of inquiry and on any given issue c can only be improved up to a point. Every area has its own natural resistance barrier. And of course the matter of affordability prevents its achievement. The overall situation stands as per Figure 2.3.

Idealization: Generality Through Loss of Detail

There are a great many *real* apples on the fruit at of the nearby supermarket. And—so the story has it—there are lots of *fictional* apples in the orchard of the

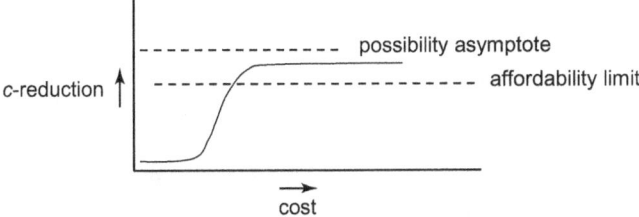

Figure 2.3: The Cost of Merit

Green Gables farm on Prince Edward Island. But in between the real and the fictional, there is also a category of apples that are neither the one nor yet quite the other neither actually real nor merely fictional. These would include such contrastively idealized things as *typical* or *ideal* Macintosh Apples.

Only with *theoretical idealizations* can we hope to secure information that is at once secure and precise: with *reality* we cannot expect to realize this combination. As Plato already saw philosophy's cognitive aspirations can only be met at the level of the ideal.

And this is a situation arises with respect to thing-kinds in general. For with such a kind Z, we can project the idea of a typical Z, a normal Z, an average Z, a perfect Z, or an optimal Z. And while we understand such references fairly well, we also realize that we may never actually encounter them in the real world. In this sense there are items whose contrastive characterization within their group is a mode of quasi-fictive idealization. And it is with such potential existents that float indecisively between the actually real and the conjecturally fictional that the present discussion will be concerned. Just what is the status of such objects and what is their place in the ontological scheme of things?

Machinery

Some clarification of concepts is needed from the outset. We shall understand a *kind* of thing as theorists have done since Greek antiquity, namely either as a natural kind (an Indian elephant) or as an artificial kind (a Y-hallmarked silver spoon). And kinds of either sort generally disaggregate into sub-kinds and eventually into ultimate kinds (*infima species*) and then into varieties, as per (for example) dog, hunting dog, beagle male adult beagle.

Moreover, our deliberations will have to resort to a certain amount of expository formalism. In specific we will need to resort to the following ideas:

- The conception of a classificatory *thing-kind* (Z, Z_1, Z_2, etc.) be it natural (corgis) or artificial (recliner chairs);
- The conception of a *descriptive property* of an item in contrast to a *non-descriptive* (and frequently *evaluative*) property. (Being *lame* is a descriptive property of a dachshund, being *cute* or *lovable* is not.);
- The conception of a *contrastive qualifier* (Q, Q_1, Q_2, etc.) for items of a thing-kind, such as an *ordinary Z*, a *normal Z*, an *unusual Z*, an *average Z*, or the like.

The descriptive properties of kind-members always indicate a determinable characteristic of some sort such as being snub-nosed, long-lived, or tawny-colored. The possession of such properties can in general be determined by inspection of the item at issue in isolation. By contrast, non-descriptive features cannot be determined from the descriptive constitution of the item. Examples of such features would be a vase's being "owned-by Napoleon" or a painting's being "inspired by Renoir."

With many contrastive qualifiers for kind-members the situation will turn out to be *quasi-fictional* in that the realization of the putative item is problematic—such a thing may or may not actually exist. For a given kind may simply fail to contain any typical or normal member whatsoever. The idea of such Q-mode Z may in the end prove to be an unrealized fiction. Let us consider such situations more closely.

Truth and Fact

There are two very different sorts cognitive failure or incorrectness. One is inadequation or discord between appearance (i.e. what we take to be so) and reality (i.e., what actually is so): namely *error* (outright mistake). The other is *inaccuracy:* the lack of exactness or precision. If the tree is 80 feet high, it is certainly false to claim it to be 95 feet, but still imprecise and inaccurate to claim it to be roughly (or approximately) 82 feet.

The following considerations are in order:
1. As an existential fact a tree cannot be approximately this or roughly that. In reality; it has to be something exact and definite;
2. But it is nevertheless an objective truth that the tree is approximately 82 feet. And this is so objectively and actually, independent of any ideas, wants, and beliefs people might have in the matter;

3. So fact and truth do not exactly correspond. And the tree is "approximately 80 feet high" but *not* adequate to reality. These need be no exact acceptance of truth with the facts;
4. Approximation defeats the classic theme that truth is coordination with fact (adequatio ad rem).

Quasi- Fictions

Combining Precision and Security:

The Typical

To implement the idea of a typicality let us take dogs as an example. There just is no typical dog as such, across the entire range from mastiff to chihuahuas. If we want typicality we have to go to something as specific as a mature female corgi. Accordingly, to capture the idea of typicality one would have to say something along the lines of the following specification.

> Given a taxonomically ultimate kind Z, a *typical* or *paradigmatic* Z would be one that (a) has all the descriptive properties that Z's generally [i.e., almost always] have, and (b) has no descriptive properties that would mark it an eccentric "odd man out" among the Z's (in that only very few of them have this property).

The need for the limitation to *descriptive* properties in this formula should be clear. Thus location, although a property unique to any given *t*, would not impede typicality since it is relational rather than descriptive.

The Normal

A normal Z is one that (1) has *almost all* of the descriptive properties that Z's almost always have, and (2) lacks *almost all* those descriptive properties that Z's almost always lack.

Given this understanding of the matter, it follows that while a typical Z will always qualify as normal as well, the reverse is not the case.

The Ordinary/Usual

An ordinary (or run-of-the-mill) Z is one that tracks the statistical norms in that it has all of the descriptive properties that most X's possess and that lacks most of the properties that most Zs lack. However, it is theoretically possible that no actual Z whatsoever is ordinary—i.e., that all the Zs there are non-ordinary in some way or other. More common of course would be that situation which all of the Zs there actually are of the usual sort. In that event, however, we will be able to say things like: If there were a Z that has the property F (e. g., "If there were a dachshund that has two tails") it would be most unusual one.

The Average

Consider the idea of an *average* Macintosh apple. To qualify as such, the apple would need to be average in size, in weight, in sugar content, in number of seeds, in red skin-coverage, etc.—with all measurable descriptive properties included. For it is of course more than likely that no actual apple would fill this bill across the board. Even when there are a great many Zs, it is easily possible that no average Z actually exists—if this is to call for averaging out in all measurable features.

And in fact they need not even always be possible. For even average Zs can prove to be impossible. Thus consider a group of three rectangles (p_1, p_2, p_3) answering to the description:

	R_1	R_2	R_3	Average
Base	3	4	5	4
Height	3	2	4	3
Area	9	8	20	12 1/3

Note that, in this particular group, a triangle that is average in base and height cannot possibly also be average in area. The concept can be applied only to particular respects, but has no aggregative applicability within the range of alternatives at issue.

Average parts need not combine to yield average wholes. Nor need ideal parts yield ideal wholes or imperfect parts yield imperfect wholes. The reality of features is complex. (Round dots need not constitute round wholes.)

On Perfection and Its Exclusion by Merit Complementarity

A perfect (or ideal) Z is one that has in maximal degree all of the positive features that can characterize Zs, and has in zero degree all of the negative features that might do so. Such things are certainly possible. Thus a coinage mint can proudly claim to have produced perfect pennies all day long. Indeed it transpires that certain (generally abstract) thing-kinds require perfection. (There are really no imperfect circles.) However, other thing-kinds by nature preclude it.

To illustrate this, consider the evaluation problem posed by the quest for a suitable dwelling. Here aspectival fission leaps to the fore. Consider such factors as size and placement. In point of size, spaciousness affords more room, is more impressive, more amenable to entertaining; but comportness small is more intimate, easier to maintain, less costly to heat, etc. And as regards placement, being in town is more accessible to work, shop-ping, whereas location in the suburbs means less traffic, more neighborliness, etc. Here and elsewhere, the various parameters of merit compete with one another in their being an evaluation.

Desideratum complementarity thus arises when two (or more) parameters of merit are linked (be it through a nature-imposed or a conceptually mandated interrelationship) in a see-saw or teeter-totter interconnection where more of the one automatically ensures less of the other, as per the situation of Figure 2.4. Here synoptic maximization is unrealizable, optimization requires compromise and prioritization.

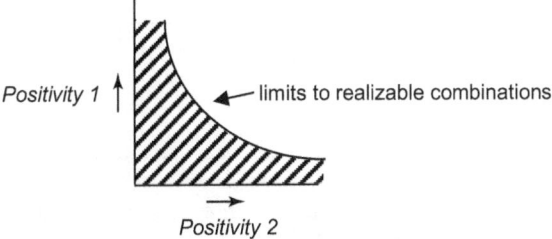

Figure 2.4: Desideratum Complementarity

Throughout such complementarity cases we have the situation that to all intents and purposes realizing more of one desideratum entail a correlative decrease in the other. We cannot have it both ways, so that ideal of achieving the *absolute perfection* at issue with a concurrent maximization of every parameter of merit at one and the same time lies outside the range of realizability as a matter of principle.

Optimality Assessment Issues

The situation arising in cases of this sort is readily illustrated. Consider a problem of evaluation with just how modes of positivity as parameters of merit, again P_1 and P_2. We assume that each can be present in three degrees: H, M, L for High, Middling, and Low respectively. The combination thus stand as per:

		P_1		
		H	M	L
	H	H	(2)	(3)
P_2	M	(4)	M	(5)
	L	(6)	(7)	L

Assessing the priority of combinations here calls for entering letters (H, M, L) in place of numbers. As Table 2.1 exhibits, we have four principal choices.

The difficulty arises where there is fixed input-determinate that is when a combination such as M/L should sometimes result in the M and sometimes in L "depending on circumstances." When there is no stable determination but there is also a matter contextual/situational instability, there will be no fixed way for resolving optimality assessment issues and evaluation requires more problematic complexities.

Table 2.1: Grading Combination

	(2)	(3)	(4)	(5)	(6)	(7)
P_1 dominance	H	H	H	M	L	L
P_2 dominance	M	L	M	L	L	L
Max-prevails	H	H	H	M	H	M
Min-prevails	M	L	M	L	L	L

And so when the various modes of merit that compare an overall desideratum are complementary, the existence of a uniquely best of them—may nevertheless be unachievable for three very different reasons:
1. There may possibly not be a single best because all of those Zs are just plain bad, with none of them any good at all. (There is no such thing as an optimal catastrophe.)
2. There may possibly not be a single best because there is a plurality of them qualified to the same optimal extent.
3. There may possibly not be a single best because no matter how good some Z is there is always another yet better.

In all of these different ways the goal of evaluative optimality may be unachievable within the range of alternatives afforded by a given kind Z.

All the same, there may be circumstances in which, even though no optimal Z exists, we may yet be able to say what an optimal Z would be like if there were one. A Z-type item which lacks to the greatest extent all of the shortcomings that actual Zs have would be indicated. Thus while none of the students in a course produce an optimal result on the examination an instructor can provide the graders with a model exam paper that realizes the desiderata of optimality—and will accordingly prove to be useful to them in their mission of grading the actual results.

The Use of Idealizations

Each of the various of quasi-idealization at issue here such as typical, normal, or average instructs of kind-exemplification relates to items which, for aught that we know to the contrary, may well not exist. The items at issue purport a descriptive specificity that can preclude its actual realization. At bottom such things are creatures of thought (*entia rationis*) rather than furnishings of the real world. But all the same, even if there is no typical Z, no average Z, or no ordinary Z, yet the contemplation of such an item puts into possession of information about the nature of reality. Even if no pupil in the class is an average performer, still knowing what such a pupil would be like provides valuable information. Irrespective of its realization the nature of such a T-type Z conveys a good deal of information about the composition of the entire group of actual Zs at issue.

And so despite their quasi-fictionality such idealization can serve a useful communicative purpose in that they can function in ways that transmits useful information. We know a great deal about the real animals at issue when we characterize a typical beagle or bulldog.

In sum, those quasi-fictional idealizations are a useful communicative resources able to render good service in descriptive and evaluative matters. The items at issue may not exist as such, but like the North Pole or the Equator they are pragmatically useful thought-instrumentalities that facilitate our proceedings within the real world. But most things are not mere fictions. Rather they are thought-instrumentalities that constitute our management of real-world issues. Their unrealizable ideality does not impede their utility in communicating about the real. Irrespective of their possible irreality as such, their employment can be highly instructive and informative aids to understanding and managing the world's actualities. Without idealizations of this "unrealistic" sort an understanding of the real and our effective operations within its orbit would be direly impeded.

The Lesson for Philosophizing

The more informative, significant, and far-ranging a philosophical generalization is in its range or the more it stands a need of adjustment qualification and limitation.

The questions of philosophizing are fundamental, complex, and being highly general are multi-aspectual, with exceptions thereby generalizes at every corner. Then as a branch of rational inquiry, the field accordingly has a high c-value. Establishing something that is at once informative, detailed, and convincing is bound to be difficult. The importance of its issue makes the field compelling. Their inherently problematic nature makes its pursuit frustrating. By the very nature of its characteristic mission the field is difficult.

From the very dawn of the subject in Greek antiquity, philosophy has taken mathematics (namely geometry) as its model—and accordingly aimed at the universality and precision that characterize this discipline. When philosophers say something along the lines of "To be real is to be part of the processes rooted in space and time." They do not intend to say that this is so usually, normally, and for the most part, but rather that it is so unqualifiedly and invariably. Strict universality and exact precision is integral to the characterizing aspirations of philosophy—and the reason why its aspirations are virtually impossible to realize—lie in the fact that within that range of the definite philosophical issues these objections really cannot be realized. Philosophy is caught in a difficult and inherently frustrating situation. For it demands conjoint precision and generality where claims can only be staked with assurance for vague and impressive approximants.

3 Contextual Acceptance and Cognitive Compartmentalization

Assertor Warrant

Opposing the ambitious claims of philosophers and theorists to the pursuit and acquisition of "the (absolute) truth," John Dewey sought to abandon the matter of truth for a more realistic and realizable conception of *warranted assertability*. This was to be something less demanding, more workable and pragmatic. But he never developed this idea vary far—and certainly not far enough to resolve the substantial logical difficulties that lie in its way.

It worried Dewey that appropriately claiming something to be true requires decisive substantiation. "I know it is true but acknowledge having no decisive grounds for its being so," looks to be a contradiction in terms. By contrast, merely to claim something as warrentedly assertable one need not go so unrealistically far. Here it would seem that three conditions have to be satisfied, namely that the contention in question must be:
- evidentially probable;
- presumptively warranted;
- circumstantially unproblematic (i.e., unimpeded by cogent counter-indications).

Correct truth claims require proof positive; by contrast, warrant—sensibly understood—merely requires probability, plausibility, and the absence of counterevidence. Such a lowering in the standard of what is required clearly enlarges the range of authorized credence.

And so Dewey and company were clearly on the right track. The truth is something so demanding that its assured achievement is almost always problematic. And so that by limiting ourselves to assured truth claims we impose drastic limits upon the range of our vision. The requisite substantiation is all too often unavailable, and many of our questions will then have to go unanswered. Accordingly, the downside of limiting oneself to "the truth, the whole truth, and nothing but the truth" is thus an impoverishment of information. The shift to a less demanding warrant overcomes this limitation.

But now, unfortunately, the prospect looms that rather than having too fewer answers we have too many on our hands, with the range of assertability growing too big for comfort. For mere warrant outruns the limit of consistency. We are embarrassed by an information glut. As in the courtroom, the inconsistency of conflicting claims confronts us on all sides. In abandoning consistency the shift from truth to warrant it creates the cognitive nightmare of allowing cases where both p and not-p have a place in the realm of acceptance.

Zones of Acceptability

The reductive epistemology of an "innocent until proven guilty" approach requires a means of ensuing consistency. And the theoretically most straightforward approach to inconsistency mitigation is to divide the overall realm of one's cognitive commitments, partitioning it into separate regions or zones of acceptance in such a way as to keep incompatibilities apart.

Of course this too has its problems. For how can one implement this fragmentation of acceptance? Is not such a position one of cognitive schizophrenia in adopting the view so vividly rejected by Pascal that truth is arbitrary, and that what counts as true on one side of the Pyrrhenies qualifies equally as false on the other?

After all, it make no sense to take a fast-and-loose approach that is indifferent to substantiation, evidentiation, and systematization. What is called for instead is the nuanced approach of a *rational contextualism*—i.e. a doctrine to the effect that what is to be acceptable as cogent and valid in one context of deliberations may not so qualify in another. On this basis acceptability is not absolute but contextual in its bearing.

And such a theory is in fact attuned to the reality of our cognitive situation. Information sources and theory complexes have varying degrees of reliability in different contexts. In some matters we entrust our credence to X and in others to Y and we sensibly proceed with context variable doctrinal system as with:
- Chinese vs. Western medicine;
- Domains of psychosomatic influence. (E.g. autosuggestion or Mesmeric effects: well developed in practice but outside the range of accepted theory;
- Orthodox vs. unorthodox modes of human behavior—e.g. hypnosis, acupuncture—allowing different discourses in different contexts.

As such illustrations suggest, implementing the conception of context-variable standards of warrant for acceptance calls for dividing and separating our claims into different subdomains of assertion because the indifferentiated overall domain of acceptance or assertion is inconsistent. The totality of cognition thus becomes partitioned into distinct subdomains or acceptance regions of such a sort that some claims of each are incompatible with the totality of the claims of every other. So unlike the domain realm of 'established truth" which must, as such, be self-consistent overall, the realm of assertion is inconsistent overall but compartmentalized into consistent subregions. On this basis one is able to speak of the truth according to this or that particular (sub)domain of deliberation with the only claims now deemed absolutely and unqualifiedly true being those that figure universally in the acceptances of each and every (sub)domain.

Contextualism

The crux is that contexts of deliberations can be separate and self-contained in nature, and that, being delimited, operations within them do not overlap but stand apart. There is no inconsistency involved in holding that "In context C_1 p holds and in context C_2 not-p." (Any more than there is in saying that "at place-time No. 1 it rains but at No. 2 it is bone dry.") And so as long as one functions *within* a context, one is consistent and it is only *across* contexts that there is inconsistency.

But which context is the correct (true, uniquely appropriate) one. The response is to reject the question as improper and illegitimate, based on a false preoccupation. It would be like asking "what is the correct tool, a hammer or a screwdriver?" Here too there is no single all-purpose answer—it all depends on the *specific* task at hand, whether to drive in a nail or a screw. The crucial point is that such a view does not give up on "real truth" or "actual correctness"; it only contextualizes it.

Just here is where the idea of Averroism comes into it. It has been widely maintained that it was the teaching of the so-called "Averrosits" of the Middle Ages that truth is divided into sectors—in particular the sectors of theological (religious) and theoretical (philosopho-scientific) truth—and that what holds good in the one domain does not necessarily do so other, so that truth was in principle plural and domain relative.[3] This closely parallels with what is now being said about assertability/acceptability.

But just what is it that prevents the contextual variation of acceptability from disintegrating into an indifferentist relativism? At this point the prospect of different doctrinal stances becomes paramount. For in this regard we must distinguish:
1. *Contextualism:* Different contexts of inquiry have different purposes in view, different ranges of application, implementations, and utilization;

---

[3] In the main this duality related to the truth of secular philosophy (especially that of Aristotle) on the one hand, and the truth religious teaching on the other. The pivotal bone of contention was the age of the universe—eternal and uncreated according to Aristotle, but created in according to Mediterranean religiosity.

2. *Functional Differentialism:* Whether or not a certain corpus of accepted information serves the purpose of a certain range of operation and effectively meets the needs of the situation obtaining there is a matter of objective experience rather than subjective instantiation;
3. *Objectivity:* The fact that the claims of a certain acceptance domain meets the needs of a certain problem area—objective and affords in impersonally cogent quality-control arbiter of cognitive adequacy.

Thus as regards validation we resort in effect to a mode of pragmatism—a justification of acceptance on a contextual basis of what succeeds in matters of issue-resolution within the relevant range of cognitive and applicative practice. For as Aristotle observed long ago "It is the mark of a knowledgeable person to expect precision in each domain only insofar as the nature of the subject admits" (*Nicomachean Ethics* 1094b 24–25).

Philosophy

Throughout its history, philosophy has pursued a quest for the truth of things. But this is ambiguous, since "the truth" at issue can be construed as:
- the certain, certifiably correct and unquestionable truth;
- the plausible and likely truth as best we can estimate it.

And here we confront the crucial question of just how high are we to set the bar of certainty and categorical assurance. And just here Aristotle's observation provides an essential admonition. It is a sobering reminder that it is this area we must be prepared to go empty handed if we are not to be linked with plausibility and reasonable assurance.

Inquiry as the circumstances of our human conditions enables us to pursue it affords us an imperfect resource. The warranted acceptability is provides no guarantee of correctness or even of answering. If we accept what is plausible and probable (rather than categorically assured) we have to curtail and compatibilize our commitments in a domain-relative way. The definitive turn is (presumably) consonant and consistent across the board. But the realm of the plausible/probable/presumptive is not.

The hunter does not always return with the prey. In philosophy we may seek for definite knowledge to resolve our concern for "All the big questions." But what we secure is generally no more than plausible conjecture.

4 Philosophical Methodology

Truth-Estimative Conjecture

Philosophers generally pursue their mission of grappling with the traditional "big questions" regarding ourselves, the world, and our place within its scheme of things by means of what is perhaps best characterized as *rational conjecture*. Conjecture comes into it because those questions arise most pressingly where the available information does not suffice—where they are not straight-forwardly answerable in terms of what has already been established. What is needed here is an *ampliative* methodology of inquiry—one that is so in C. S. Peirce's sense of propounding contentions whose assertoric content goes beyond the evidence in hand.[4] We need to do the very best we can to resolve questions that transcend accreted experience and outrun the reach of the information already at our disposal. It thus becomes necessary to have a way for obtaining the best available, the "rationally optimal," answers to our information-in-hand-transcending questions about how matters stand in the world. And experience-based conjecture—theorizing if you will—is the most promising available instrument for question-resolution in the face of imperfect information. It is a tool for use by finite intelligences, providing them not with the best *possible* answer (in some rarified sense of this term), but with the best *available* answer, the putative best that one can manage to secure in the actually existing conditions in which we do and must conduct our epistemic labors.

Notwithstanding, those guarding caveats, the "best available" answer at issue here is intended in a rather strong sense. We want not just an "answer" of some sort, but a viable and acceptable answer—one to whose tenability we are willing to commit ourselves. The rational conjecture at issue is not to be a matter of *mere guesswork*, but one of *responsible estimation* in a strict sense of the term. It is not *just* an estimate of the true answer that we want, but an estimate that is sensible and defensible: *tenable*, in short. We may need to resort to more information than is actually given, but we do not want to make it up "out of thin air." The provision of reasonable warrant for rational assurance is the object of the enterprise.

---

[4] For Charles Sanders Peirce, "ampliative" reasoning is synthetic in that its conclusion goes beyond ("transcends") the information stipulated in the given premises (i.e., cannot be derived from them by logical processes of deduction alone), so that it "follows" from them only inconclusively. Cf. Peirce 1931, sections. 2.680 and *passim*.

In the information-deficient, enthymematic circumstances that prevail when questions must be resolved in the face of evidential underdetermination, we have and can have no logically airtight *guarantee* that the "best available" answer is actually true. Given that such truth-estimation involves transcending the information at hand, we know that rational inference cannot guarantee the truth of its products. (Indeed, if the history of human inquiry has taught us any one thing, it is the disastrous metainduction that the best estimate of the truth that we can make at any stage of the cognitive game all too frequently comes to be seen to be well off the mark with the wisdom of eventual hindsight.) Rational inquiry is a matter of doing no more—but also no less—than the best we can manage to realize in its prevailing epistemic circumstances. Nevertheless, the fact remains that the rationally indicated answer does in fact afford our most promising *estimate* of the true answer—that for whose acceptance as true the optimal overall case be constructed in the circumstances at hand.

The need for such an estimative approach is easy to see. After all, we humans live in a world not of our making where we must do the best we can with the limited means at our disposal. We must recognize that there is no prospect of assessing the truth—or presumptive truth—of claims (be they philosophical or scientific) independently of the use of our imperfect mechanisms of inquiry and systematization. And here it is *estimation* that affords the best means for doing the job. We are not—and presumably will never be—in a position to make totally secure claims to the definitive truth regarding those great issues of philosophical interest. But we certainly can—and indeed must—do the best we can to achieve a reasonable *estimate* of the truth. We can and do *aim* at the truth in our inquiries even in circumstances where we cannot make failproof pretensions to its attainment, and where we have no alternative but to settle for the *best available estimate* of the truth of the matter—that estimate for which the best case can be made out according to the appropriate standards of rational cogency. And *systematization* in the context of the available background information is nothing other than the process for making out this rationally best case. It is thus rational conjecture based on systematic considerations that is the key method of philosophical inquiry, affording our best hope for obtaining promising answers to the questions that confront us.

The Problem of Data

It is informative and interesting to approach a philosophical text from the angle of the question of *authority*, and to ask ourselves, line by line and claim by claim: On what sort of basis can the author expect us to accept the assertion at issue? Is it as a matter of scientific fact, of common sense—of "what everybody should realize," of accepting the assertion of some expert or authority, of intuitive self-evidence, of drawing a suitable conclusion from previously established facts, or just what? Ultimately, the issue of acceptability is always one of considerations we are expected to endorse or concede because of the plausibility of their *credentials*. And this has many ramifications.

Neither individually nor collectively do we humans begin our cognitive quest empty handed, with a tabula rasa. Be it as single individuals or as entire generations, we always start our inquiries—even in philosophy—with the benefit of a diversified cognitive heritage, falling heir to that great mass of information and misinformation that is the "accumulated wisdom" of our predecessors—or those among them to whom we choose to listen. What William James called our "funded experience" of the world's ways—of its nature and our place within it—constitute the *data* at philosophy's disposal in its endeavor to accomplish its question-resolving work. These "data" of philosophy include:

1. Common-sense beliefs, common knowledge, and what have been "the ordinary convictions of the plain man" since time immemorial;
2. The facts (or purported facts) afforded by the science of the day, the views of well-informed experts and authorities;
3. The lessons we derive from our dealings with the world in everyday life;
4. The received opinions that constitute the worldview of the day; views that accord with the "spirit of the times" and the ambient convictions characteristic of one's cultural heritage;
5. Tradition, inherited lore, and traditionary wisdom (including religious tradition);
6. The "teachings of history" as best we can discern them.

No plausible source of information about the world and our place within it fails to bring grist to philosophy's mill. The whole range of the (purportedly) established "facts of experience" furnishes the extra-philosophical inputs for our philosophizing—the materials, as it were, for our philosophical reflections.

All of philosophy's data deserve respect: common sense, tradition, general belief, accepted (i.e. well established) prior theorizing—the sum total of the different sectors of "our experience." They are all plausible, exerting some degree of cognitive pressure and having some claim upon us. They may not constitute irrefutably established knowledge, but nevertheless they do have some degree of merit and, given our cognitive situation, it would be very convenient if they turned out to be true. The philosopher cannot simply turn his back on these data without further ado. Still, even considering all this, there is nothing sacred and sacrosanct about the data. For, taken as a whole, the data are too much for tenability—collectively they run into conflicts and contradictions. The long and short of it is that the data of philosophy constitute a plethora of fact (or purported fact) so ample as to threaten to sink any ship that carries so heavy a cargo. For those data are by no means unproblematic. The constraint they put upon us is not peremptory and absolute—they do not represent certainties to which we must cling at all costs. What we owe to these data, in the final analysis, is not *acceptance* but merely *respect*. Even the plainest of "plain facts" can be questioned, as indeed some of them must be, seeing that, in the aggregate, they are collectively inconsistent.

Philosophizing accordingly roots in contradiction—in conflicting belief-tendencies. Philosophical problems arise in a cognitive setting, not wholly of our making, that is rationally intolerable; the overall aggregate of contentions we deem plausible involves us in logical inconsistencies. The cognitive situation is always deeply problematic in its initial and presystemic state. The impetus to philosophizing arises when we step back to look critically at what we know (or *think* we know) about the world and try to make sense of it. We want an account that can optimally accommodate the data—recognizing that it cannot, in the end, accept them all at face value. Philosophy does not furnish us with new ground-level facts; it endeavors to systematize, harmonize, and coordinate the old into coherent structures in whose terms we can meaningfully address our larger questions. The prime mover of philosophizing is the urge to systemic adequacy—to achieving consistency, coherence, and rational order within the framework of what we accept. Its work is a matter of the *disciplining* of our cognitive commitments in order to make overall sense of them—to render them harmonious and coherent. And so the demands of rational consistency come to the forefront.[5]

Two injunctions regarding the mission of rational inquiry set the stage for philosophy:

> Answer the questions! Say enough to satisfy your need for information about things.
> Keep your commitments consistent! Don't say so much that some of your contentions are in conflict with others.

There is a tension between these two imperatives—between the factors of commitment and consistency. We find ourselves in the discomfiting situation of cognitive conflict, with different tendencies of thought pulling in divergent direc-

---

[5] This view of philosophy accords closely with the spirit of Aristotle's description of the enterprise in the opening section of book beta of the *Metaphysics*, with its stress on the centrality of apories.

tions. The task is to make sense of our discordant cognitive commitments and to impart coherence and unity to them insofar as possible.

Here, then, we come to one of the core issues of the domain. Philosophizing on any topic always moves through two stages. At first, there is a "presystemic" stage, where we confront a group of tentative commitments, all viewed as more or less acceptable, but which are collectively untenable because of their incompatibility. Subsequently, there comes a "systematizing" phase of facing up to the inconsistency of the raw material represented by the "data." And this becomes a matter of eliminative pruning and tidying up where our commitments have been curtailed to the point where consistency has been restored. Those "data" of philosophy are invariably the deliverances of fallible sources of information that afford misinformation as well, so that the process of getting our answers to fit with the data also involves smoothing out the data themselves.

The key task of philosophy is thus to impart systemic order into the manifold of relevant data; to render them coherent, harmonious, and, above all, consistent. One might, in fact, define philosophy as the rational systematization of our beliefs on the issues—the fundamentals of our understanding of the world and our place within it. We become involved in philosophy in our endeavor to make systematic sense of the extra-philosophical "facts"—when we try to answer those big questions by systematizing what we think we know about the world, pushing our "knowledge" to its ultimate conclusions and combining items usually kept in convenient separation. Philosophy is the policeman of thought, as it were, the agent for maintaining law and order in our cognitive endeavors. Its task is to render our "experience" (in the broadest sense of the term) cogent and intelligible.[6]

Why Not Simply "Live With Inconsistency"? The Imperative of Cognitive Rationality

The pursuit of rational coherence—consistency, compatibility, comprehensiveness—is the crux of philosophical method. But is this emphasis on structure, order, and logical elegance in fact justified? Is systematic coherence and consistency itself not simply the hobgoblin of small minds. Is it not a mere ornament—a dispensable luxury?

To Alice's insistence that "one can't believe impossible things," the White Queen replied: "I daresay you haven't had much practice. When I was your

---

[6] These aspects of philosophizing are explored at greater length in Rescher 1985.

age, I always did it for half-an-hour a day. Why, sometimes I've believed as many as six impossible things before breakfast." But even with practice, the task is uncomfortable and unsatisfying. A profound commitment to the demands of rationality is a thread that runs through the whole fabric of our philosophizing; the dedication to consistency is the most fundamental imperative of reason. "Keep your commitments consistent" is philosophy's ruling injunction. Inconsistent claims are utterly uninformative. We don't want just answers, but reasoned answers, defensible answers that square with what we are going to say in other contexts and on other occasions. And this means that we must go back and clean out the Augean stable of our cognitive inclinations, seeing that the commitment to rational coherence is a part of what makes philosophy the enterprise it is.

Yet is consistency itself something altogether fixed and definite? What of the fact that there are different systems of logic? Does this not open up the prospect that one thinker's inconsistency is another's compatibility? Perhaps so. But at this point we must be maximally strict. If even the most fastidious logician discerns problems, we must undertake to worry. In the interests of philosophical adequacy, the propositions we juxtapose must, like Caesar's wife, be above suspicion; if there is *any* plausible basis for charges of incompatibility in *any* viable system of logic, then adjustments are in order.

It must be emphasized that the impetus to rationality does not in any way prejudge the *outcome* of our theorizing. It may well turn out in the end that the "principle of noncontradiction" does not hold of the world; reality as best we can discern it may possibly turn out to be inconsistent. But what is presently at issue is not reality as such, but our *account* of it. Regardless of the world's consistency, our *theory* of it must be self-consistent if it is to merit serious consideration. And here it is important to recognize that thought need not necessarily share the features of its object. A sober study of inebriation is perfectly possible, as is a coherent characterization of the opinions of an incoherent thinker or a consistent characterization of an inconsistent system (where we insert another assertor—perhaps "the nature of things"—between ourselves and those "inconsistent facts"). A coherent theory of an inconsistent reality can perfectly well be contemplated.[7] A methodological insistence on consistency does not prejudge the ontological nature of the real; what is at issue is simply the consistency and coherence of our own deliberations. We might in the end be driven by rational considerations to accept the conclusion that reality is inconsistent, but this is

---

[7] To be sure, its details must be wrapped in the intricacies of semantical theory. See Rescher and Brandom 1979.

no reason to cease striving for consistency in our *theory* of reality—at any rate until such time as a clear demonstration of the actual impossibility of reaching this goal becomes available.

After all, to endorse a discordant diversity of claims is in the end not to enrich one's position through a particularly generous policy of acceptance, but to impoverish it. To refuse to discriminate is to go empty handed, without answers to our questions. It is not a particularly elevated way of doing philosophy—but a way of not doing philosophy at all. For it evades the problems of the field, abandoning the traditional project of philosophy as rational problem solving. We are compelled to systematize our knowledge into a coherent whole by regimenting what we accept in the light of principles of rationality. Philosophizing is a work of reason; we want our problem resolutions to be backed by good reasons—reasons whose bearing will doubtless not be absolute and definitive but will, at any rate, be as compelling as is possible in the circumstances. Reasoning and argumentation are thus the life blood of philosophy. If we do not have a doctrine that is consistent and coherent, then we have nothing.

Of course, no rational guarantee of categorical assurance can be issued in advance to show that our philosophical efforts at systematizing our knowledge of the world is bound to succeed. The systematicity of our knowledge is (as we shall see) not something that can be guaranteed *a priori*, as having to obtain on the basis of the "general principles" of the matter. The parameters of harmonious systematicity—coherence, consistency, uniformity, and the rest—represent a family of *regulative ideals* towards whose realization our cognitive endeavors do and should strive. But this impetus to systematicity is the operative expression of a governing ideal, and not something whose realization can be taken for granted as already certain and settled from the very outset. The extent to which our efforts at philosophy can manage to succeed in achieving the objectives at issue is always "something that remains to be seen"—in *this* regard replicating exactly the situation of natural science.

Systematicity and the Impetus to Coherence

Philosophizing is a matter of bringing question-answering commitments into alignment with the varied and often dissonant data of experience. But just how can such a process be expected to work?

The coherentist approach to rational substantiation proceeds by way of a network model that sees a cognitive system as a family of inter-related theses, not necessarily arranged in a *hierarchical* arrangement (as with an axiomatic system), but rather linked with one another by an *interlacing network* of connections. These interconnections are *confirmatory* in nature, but not necessarily *deductive* (since the providing of "good explanatory accounts" rather than "logically conclusive grounds" is ultimately involved).

A network system dispenses with one advantageous feature that characterizes Euclidean systems *par excellence*. Since everything in a deductive system hinges upon the axioms, these will be the only elements that require any independent support or verification. Once they are secured, all else is supported by them. The upshot is a substantial economy of operation: since everything pivots about the axioms, the bulk of our epistemological attention can be confined to them. A network system, of course, lacks an axiomatic basis, and so lacks this convenient feature of having one delimited set of theses to carry the burden of the whole system upon its shoulders. On the network model, the process of justification need not proceed along a linear path. Its mode of justification is in general nonlinear, and can even proceed by way of (sufficiently large) cycles. Two very different conceptions of explanatory procedure are at issue. The Euclidean approach is geared to an underlying conception of fundamentality or logical dependency in the Aristotelian sense of *priority*, in terms of what is supposed to be "better understood." Its procedure is one of *reduction by derivation:* reducing derivative, "subservient" truths to their more fundamental "master" truths. By contrast, the network appeal is unreductive. Its motto is not "Explanation by derivation" but "Explanation by interrelation."

But even when its linkages operate along wholly deductive lines, a network model would still depart drastically from the geometric paradigm. For from the network standpoint, the classical Euclidean model imposes a drastic limitation in inflating what is at most a *local* feature of derivation from the underived (i.e. *locally* underived) into a *global* feature that endows the whole system with an axiomatic structure. What matters is that the network links theses in a complex pattern of relatedness by means of some (in principle variegated) modes of probative interconnections. The network theorist does not deny that a cognitive system must have a *structure* (how else could it be a system!). But it recognizes that this structure need not be of the form of a rank ordering—that it can provide for the more complex interrelationships that embody a reciprocity of involvement. It is no longer geared to the old hierarchical world picture that envisages a unidirectional flow of causality from fundamental to derivative orders of nature.

An important advantage of a network system over one that is axiomatic/deductive inheres in the former's accommodation of relatively self-contained subcycles. This absence of a rigidly linear hierarchical structure is a source of strength and security. In an axiomatic system a change anywhere ramifies into a change everywhere—the entire structure is affected when one of its supporting layers is removed. But with a network system that consists of an integrated organization of relatively self-sufficient components, certain of these components can be altered without dire repercussions for the whole.[8] C. S. Peirce observed this aspect of network systematization when he wrote:

> Philosophy ought to imitate the successful sciences in its methods, so far as to proceed only from tangible premises which can be subjected to careful scrutiny, and to trust rather to the multitude and variety of its arguments than to the conclusiveness of any one. Its reasoning should not form a chain which is no stronger than its weakest link, but a cable whose fibers may be ever so slender, provided they are sufficiently numerous and intimately connected.[9]

On the network-model approach to the organization of information, there is no attempt to erect the whole structure on a foundation of basic elements, and no necessity to move along some unidirectional path—from the basic to the derivative, the simple to the complex, or the like. One may think here of the contrast between the essentially linear order of an expository book, especially a textbook, and the inherently network-style ordering of an entire library or an encyclopedia. Again, the contrast between a taxonomic science (like zoology or mineralogy) and a deductive science (like classical celestial mechanics) can also help to bring out the difference between the two styles of cognitive organizations.

One vivid illustration of the network approach to organizing information come from textual interpretation and exegesis. Here there is no rigid, linear pattern to the sequence of consideration. The whole process is iterative and cyclical; one is constantly looking back to old points from new perspectives, using a process of feedback to bring new elucidations to bear retrospectively on preceding analyses. What determines correctness here is the matter of over-all fit, through which every element of the whole interlocks with some others. Nothing need be more fundamental or basic than anything else: there are no absolutely fixed pivot points about which all else revolves. One has achieved adequacy when—through a process that is continually both forward and backward looking—one has reached a juncture where everything stands in due mutual coordination with everything else. The key operative idea is that of *explanation through systematization*—i.e. solving the puzzle by "getting all the pieces to fit properly" so that a comprehensive picture emerges which "makes sense" by putting everything into place.

---

8 Compare Simon 1965, pp. 63–76
9 Peirce 1934, sect. 5.265.

Network systematization is best approached in the light of a contrast between two profoundly different approaches to the cognitive enterprise which, for want of better choices, might be called the *expansive* and the *reductive*, respectively.[10]

The *expansive* strategy searches for a suitable basis of highly secure and unproblematically acceptable propositions that are acceptable as "true beyond reasonable doubt." Given such a carefully circumscribed and tightly controlled starter-set of secure propositions, one proceeds to move outwards ampliatively by making inferences from this secure starter set. The resulting picture is that of a bull's eye target moving from the center outwards.

We here proceed by moving expansively outward from the secure home base of an entirely unproblematic core. This is the classical foundationalist approach of mathematics. It will not do in philosophy. Here we must look elsewhere—to coherentism.

By contrast with the expansive approach of traditional foundationalism, the reductive strategy proceeds in exactly the opposite direction. It abandons the quest for an initial starter-set of secure and unproblematically acceptable truths. Instead, its starting point is a relatively generous and undemanding quest for well-qualified candidates or prospects for truth. Thus at the outset, one does not require contentions that are certain and altogether qualified for recognition as genuine truths, but rather proposition that are no more than plausible, initially promising candidates for endorsement that exert an attraction on us at first view. Of course, not all of these inherently meritorious truth-candidates are to be endorsed or accepted as true, seeing that we recognize from the very outset that we cannot simply adopt the whole lot, because they are competing and conflicting—mutually contradictory. What we have to do is to impose a delimiting (and consistency-restoring) screening-out that separates the sheep from the goats until we are left with something that merits endorsement. And here we proceed by way of diminution or compression. Like a bull's eye target moving from the periphery to the center, such a contractive approach reflects a coherentism that is the inverse of the expansive approach of traditional foundationalism. For its reductive process proceeds by *narrowing* that over-ample range of plausible prospects for endorsement. The process is one of pruning and elimination by means of evaluations based on considerations of harmony and best fit—a reduction of overdeterminative data through the use of the parameters of systematization as guides to plausibility. Using guideposts like simplicity, uniformity, regu-

---

[10] Compare the treatment of data in Rescher's 1973 and his 1976, where the relevant issues are treated in considerable detail.

larity, analogy, and the like, we prime the data with a view to endowing the residually, retained data with as much order and harmony as is achievable in the circumstances. The coherence theory thus implements F. H. Bradley's thesis that *system* (i.e. systematicity) provides a test-criterion most appropriately fitted to serve as arbiter of truth.

As we have seen, it is the overdeterminative situation of a collective inconsistency of "the data" that characterizes the problem situation of philosophy, so that the coherentist methodology affords the most natural approach to the epistemology of the field. The process of deriving significant and consistent results from an inconsistent body of information is a key feature of the coherence theory of truth, which faces (rather than, with standard logic, evades) the question of what inferences can appropriately be drawn from an inconsistent body of information. The initial mass of inconsistent information are the data for applying the mechanism of coherence as a criterion of truth, and its product is a consistent system of acceptable truths. Such an approach assumes an entirely *inward* orientation: it does not seek to compare the truth candidates directly with "the facts" obtaining outside the epistemic context; rather, having gathered in as much information (and this will include also *misinformation*) about the facts as possible, it seeks to sift the true from the false *within* this body.

The coherentist approach unhesitatingly espouses the principle that knowledge is "true, justified belief," subject, however, to construing this as justifying warrant at issue is provided by *an appropriate sort of systematization*. However, since the systematization at issue is viewed as being of the network type, the impact of the thesis is drastically altered. For we now envisage a variant view of justification, one which radically reorients the thesis from the direction of the foundationalists' quest for an ultimate basis for knowledge as a quasi-axiomatic structure. Now "justified" comes to mean not "derived from basic (or axiomatic) knowledge," but rather "appropriately interconnected with the rest of what is known." Philosophizing thus consists in a rational rebuilding of the structure of our beliefs in the effort to do what we can to erect a solid and secure structure out of the ill resolved contents placed at our disposal by our initial restrictions to belief. On this approach, the validation of an item of knowledge—the rationalization of its inclusion alongside others within "the body of our knowledge"—proceeds by way of exhibiting its interrelationships with the rest: they must all be linked together in a connected, mutually supportive way (rather than having the form of an inferential structure built up upon a footing of rock-bottom axioms).

Philosophizing As the Conjectural Systematization of Experience

While the paradigm procedure for ampliative reasoning is *plausibilistic and inductive*, the paradigm procedure for contraction is *dialectical argu-mentation*. To effect the necessary reductions we do not proceed via a single inferential chain, but through backing and filling along complex cycles of reasoning which criss-cross over the same ground from different angles of approach in their efforts to identify and eliminate the weak spots. The object of the exercise is to determine how smoothly and harmoniously a thesis can be enmeshed in the overall fabric of diverse and potentially discordant and competing contentions. We are now looking for the best candidates among competing alternatives—for that resolution for which, on balance, the strongest overall case can be made out. It is this second, reductive, approach that typifies the procedure of philosophy. Here, accordingly, it is not "the uniquely correct answer" but "the least problematic, most defensible position" that we seek. And the crux of our standard of acceptance lies, as we have seen, not with the issue of secure premises but with the issue of sensible conclusions—results that fit most smoothly and harmoniously within our overall commitment to the manifold "data" at stake in philosophical matters.

One further important point should also be stressed in this connection. To someone accustomed to thinking in terms of a sharp contrast between organizing the information already in hand and an active inquiry aimed at extending it, the idea of a *systematization of conjecture with experience* may sound like a very conservative process. This impression would be quite incorrect. Inquiry—and philosophical inquiry above all—must not be construed to slight the dynamical aspect. And systematization itself is an instrument of inquiry—a tool for aligning question-resolving conjecture with the (of itself inadequate) data at hand. The factors of completeness, comprehensiveness, inclusiveness, unity, etc. are all crucial aspects of system, and the ampler the information-base, the ampler is the prospect for our systematization to attain them. The drive to system embodies an imperative to broaden the range of our experience, to extend and expand the data-base from which our theoretical triangulations proceed. In the course of this process, it may well eventuate that our existing systematizations—however adequate they may seem at the time—are untenable and must be overthrown in the interest of constructing ampler and tighter systems. Philosophical systematization is emphatically not a blindly conservative process which only looks to what fits smoothly into heretofore established patterns, but one where the established patterns are themselves ever vulnerable and liable to be upset in the interests of devising a more comprehensive systematic framework.

A formal theory constructed by means of the systemic method must not in any way come into outright conflict or contradiction with our informal, presystematic conceptions. If such a conflict did arise, the whole purpose of the inquiry would be defeated, for we would no longer be studying the concepts which we had set out to investigate, but radically different ones. This is why the method of counter-examples is such a useful and powerful instrument in these conceptual investigations. An analysis starting from the ordinary, informal concepts of a given domain must, if acceptably executed, end with results that are fully compatible with ordinary conceptions.

However, while the systematized conceptual reworking that constitutes a theory of the kind our method is designed to provide, must not conflict with our informal conceptions, they may, and indeed should go beyond them in such important respects as precision of meaning and explicitness of logical relationships. Ernest Nagel has urged this point with characteristic cogency:

> No ... system of formal logic is or can be just a faithful transcription of those inferential canons which are embodied in common discourse, though in the construction of these systems hints may (I would say *must*) be taken from current usage; for the entire *raison d'être* for such systems is the need for precision and inclusiveness where common discourse is vague and incomplete, even if as a consequence their adoption as a regulative principles involves a modification of our inferential habits.[11]

The current popularity of analytical techniques of philosophical enquiry
Should not blind us to the usefulness and propriety of synthetical, i.e. system-constructive, techniques.

It might be urged in objection that, given the fact that our informal conceptions act as ultimate arbiter under the present method there is no point in concerning ourselves with the "different" concepts to be encountered in a formalized system based upon them. This objection can be formulated in the form of a dilemma. When the system agree: with our informal ideas, it is dispensable (since we could rely upon them alone), while when it disagrees with our intuitions, it evince its own inadequacy, since such disagreement must be taken as evident: for its incorrectness and unacceptability. The dilemma rests on a mistake. It closes its eyes to the frail, partial, and fragmentary character of our informal ideas regarding such concepts as lie within the proper sphere of application of the present method. *Ex post facto* agreement of intuition with systematically attained results is a very different thing from direct intuitive attainment of the results themselves. Our informal conceptions prove, in such

---

11 Nagel 1944, p. 205.

matters, to be only a very partial guide. They start us off in the proper direction, and they help from time to time along the way, but they do not guide us along the whole course of the journey. Here, the task of systematization becomes indispensable. Only when these are properly collected, collated, focused, systematized, and (judiciously) projected, can our informal idea: provide adequate guidance. And in this way position can be reached that could never be attained by unassisted intuition acting directly. A helpful analogy is that of mental calculation *vis à vis* longhand computations. The systematic procedures (i.e. longhand computations) must never conflict or prove to be incompatible with the informal instrumentalities (menu calculation), but they vastly extend the range of our logical and enable us to see our way clear to conclusions not within the reach of informal procedures alone.

The pragmatic theory of truth espoused by James, Dewey, and others, takes practicability as its criterion of truth, assessing truth by such questions as "Does it work?", "Can it be applied and used successfully?", "Is it viable in practice?" It is apparent that this theory simply borrows its criterion of truth from the field of methodology, i.e., the *theory of method*. And however dubious and questionable the pragmatic thesis may be as a criterion of the truth of propositions or theories, it is clearly quite correct and wholly valid within its original and proper sphere: as a criterion of the acceptability and justification of *methods*. The vindication of a method (or tool, instrumentality, procedure, practice, and the like) can clearly be based on the extent to which it *works*, and upon this criterion alone. As a doctrine regarding the criterion of evaluation for methods (etc.) pragmatism is perfectly sound, however futile and erroneous it may be regarded as a doctrine as to the truth of propositions.

When these general considerations are applied to the specific instance of the method of conceptual analysis which I have been concerned to describe, we conclude that the question of its justification ought not to be raised as a distinct and isolated issue. For if a justification is to be had, it cannot be based upon any general or abstract considerations: it must be sought in the specific applications of the method. The sole justification that can be given to this (or any other method) is one based upon realization that it stands or falls with the success of such instances of its application: its justification will have to rest upon its success in furnishing the means for solving specific, identifiable problems.

5 Aporetic Method in Philosophy

Philosophical Apories

Philosophizing may "begin in wonder," as Aristotle said, but it usually soon runs into puzzlement and perplexity. We have many and far-reaching questions about our place in the world's scheme of things and endeavor to give answers to them. But generally the answers that people incline to give to some questions are incompatible with those they incline to give to others. (We sympathize with the skeptics, but condemn the person who doubts in the face of obvious evidence that those drowning children need rescue.) We try to resolve problems in the most straightforward way. But the solutions that fit well in one place often fail to square with those that fit smoothly in another. Cognitive dissonance rears its ugly head and inconsistency arises. The impetus to remove such puzzlement and perplexity is a prime mover of philosophical innovation.

An apory is a group of contentions that are individually plausible but collectively inconsistent.[12] The things we incline to maintain issue in contradiction. One can encounter apories in many areas—ordinary life, mathematics, and science included—but they are particularly prominent in philosophy. For the wide-ranging and speculative nature of the field—the fact that it addresses questions we want to raise but almost dare not ask—means that the range of our involvements and commitments is more extensive, diversified, and complex here than elsewhere. For it lies in the nature of the field that in philosophy we must often reason from mere *plausibilities,* from tempting theses that have some substantial claim on our acceptance but are very far from certain. And so it can transpire here that the theses we endorse are inconsistent—conflicting plausibilities rather than assured compatible truths. Thus aporetic situations arise-circumstances in which the various theses we are minded to accept prove to be collectively incompatible.

Consider an historical example drawn from the Greek theory of virtue:
1. If virtuous action does not produce happiness (pleasure) then it is motivationally impotent and generally pointless;
2. Virtue in action is eminently pointful and should provide a powerfully motivating incentive;
3. Virtuous action does not always-and perhaps not even generally-produce happiness (pleasure).

---

[12] The word derives from the Greek ἀπορία on analogy with "harmony" or "melody" or indeed "analogy" itself.

It is clearly impossible-on grounds of mere logic alone-to maintain this family of contentions. At least one member of the group must be abandoned. And so we face the choice among three modes of *abandonment:*
- *Abondonment 1:* Maintain that virtue has substantial worth quite on its own account even if it does not produce happiness or pleasure (Stoicism, Epictetus, Marcus Aurelius);
- *Abondonment 2:* Dismiss virtue as ultimately unfounded and unrationalizable, viewing morality as merely a matter of the customs of the country (Sextus Empiricus) or the will of the rulers (Plato's Thrasymachus);
- *Abondonment 3:* Insist that virtuous action does indeed always yield happiness or pleasure-at any rate to the right-minded. Virtuous action is inherently pleasure-producing for fully rational agents, so that the virtue and happiness are inseparably interconnected (Plato, the Epicureans).

This illustration exemplifies the situation of an aporetic cluster: an inconsistent group of plausible contentions to which the only sensible reaction is the abandonment of one or another of them. Our cognitive sympathies have become overextended and we must make some curtailment in the fabric of our commitments. Note, moreover, that in aporetic situations, unlike elsewhere, the option of a suspension of judgment is foreclosed; the mere rejection of a thesis is tantamount to the acceptance of its negation. For suppose that A, B, C is an inconsistent triad composed of theses we deem individually plausible. Then, by the hypothesis that we are minded "to accept as much as possible," when we drop C we are in a position to accept A and B, and these (by the condition of inconsistency) entail not-C.

The Data of Philosophy and Its Aporetic Nature

As stage-setting for the discussion it is useful to take a look at the aporetic nature of philosophy.

Just what sort of things constitute "the data" of philosophy? They will clearly include:
- Common-sense beliefs, common knowledge, and what have been "the ordinary convictions of the plain man" since time immemorial;
- The facts (or purported facts) afforded by the science of the day; the views of well-informed "experts" and "authorities";
- The lessons we derive from our dealings with the world in everyday life;

- The received opinions that constitute the worldview of the day; views that accord with the "spirit of the times" and the ambient convictions of one's cultural context;
- Tradition, inherited lore, and ancestral wisdom (including religious tradition);
- The "teachings of history" as best we can discern them.

Broadly speaking, what is at issue here are "the lessons of experience" as best the resources at our disposal enable us to discern them.

All these data should be treated with consideration. The philosopher cannot simply turn his back on these data without further ado. They are all "plausible," exerting some degree of cognitive pressure and having some claim upon us. They may not constitute irrefutably established knowledge, but nevertheless they do have some degree of epistemic merit, and given our cognitive situation, it would be very convenient if they turned out to be true.

However, the fact of the matter is that these data are by no means unproblematic. For these data constitute a plethora of fact (or purported fact) so ample as to embody inner inconsistencies and contradictions. They accordingly threaten to sink any ship that carries so heavy a cargo. The difficulty is—and always has been—that the data of philosophy afford an embarrassment of riches. They are not only manifold and diversified but invariably yield discordant results due to a cognitive overcommitment within which inconsistencies arise. Taken altogether in their grand totality, the data of philosophy are inconsistent.

Philosophy accordingly roots in contradiction—in conflicting beliefs. The problems of this discipline arise in a cognitive setting, not wholly or our making, that is rationally intolerable; the overall set of contentions we deem plausible lead us into logical inconsistencies. The cognitive situation is always deeply problematic in its initial and presystemic state. The impetus to philosophizing arises when we step back to look critically at what we know (or think we know) about the world and try to make sense of it. We want an account that can optimally accommodate the data—recognizing that it cannot, in the end, accept them all at face value. The constraint they put upon us is not peremptory and absolute—they do not represent certainties to which we must cling at all costs. What we owe these data, in the final analysis, is *respect*, not *acceptance*. In particular, even the plainest of "plain facts" can be questioned.

An *aporetic cluster* is a family of philosophically relevant contentions of such a sort that: (1) as far as the known facts go, there is good reason for accepting them all; the available evidence speaks well for each and every one of them, but (2) taken together, they are mutually incompatible; the entire family is inconsistent. Such a cluster is a set of otherwise congenial propositions that, unfortunately, happen to be mutually inconsistent. They cannot all be *right*—their mutual inconsistency precludes that prospect; but they are all *plausible*, all seemingly acceptable and to some extent appealing. They put us into the unhappy position of having to make difficult cognitive choices. And philosophizing, broadly understood, calls for making such choices as best we can.

On such a perspective, philosophy does not furnish us with new ground-level facts; it endeavors to systematize and coordinate the old into coherent structures by whose means we can meaningfully address our larger questions. The prime mover to philosophizing is the urge to systemic adequacy—to bringing consistency, coherence, and rational order into the framework of what we accept. Its work is a matter of the *disciplining* of our cognitive commitments in order to make overall sense of them. And so the demands of rational consistency come to the forefront.[13] In theory we could, to be sure, simply suspend judgment in such a case and abandon the entire cluster rather than try to localize the difficulty in order "to save what we can." But this is too great a price to pay. By taking this course of wholesale abandonment we lose too much by forgoing answers to too many questions. We would thereby curtail our information not only beyond necessity but beyond comfort as well, seeing that we have some degree of commitment to all members of the cluster and do not want to abandon more of them than we have to. To obtain tenable answers to our questions in philosophy we must resolve these tensions by removing inconsistencies.

The Aporetic Perspective

Doing nothing is not a rationally viable option when we are confronted with a situation of aporetic inconsistency. Something has to give. Someone (at least) of those incompatible contentions at issue must be abandoned. Apories constitute situations of *forced choice:* an inconsistent family of theses confronts us with an unavoidable choice among alternative positions.

When one confronts an aporetic situation there are only two rationally viable alternatives: one can throw up one's hands, become a sceptic, and walk away from the entire issue, or else one can settle down to the work of problem solving, trying to salvage by way of cognitive damage control enabling one to make the best of a difficult situation. This latter course clearly has greater intellectual appeal.

---

[13] This view of philosophy accords closely with the spirit of Aristotle's description of the enterprise in the opening section of book beta of the *Metaphysics*, with its stress on the centrality of apories.

More Examples

Apories—collective inconsistency among individually plausible contentions—structure the philosophical landscape. They show how various positions are interlocked in a mutual interrelationship that does not meet the eye at first view because the areas at issue may be quite disparate.

Consider, for example, the following apory:
1. All knowledge is grounded in observation (empiricism);
2. We can only observe matters of empirical fact;
3. From empirical facts we cannot infer values (the fact-value divide);
4. Knowledge about values is possible (value cognitivism).

Given that (2) and (3) entail that value statements cannot be inferred from observations, we arrive via (1) at the denial of (4). Inconsistency is upon us. There are four ways out of this trap:

- *Rejection 1:* There is also non observational, namely, intuitive or instinctive, knowledge—specifically of matters of value (value-intuitionism; moral-sense theories);
- *Rejection 2:* Observation is not only sensory but also affective (sympathetic, empathetic). It thus can yield not only factual information but value information as well (value-sensibility theories) plausible reasoning, such as "inference to the best explanation" (values-as-fact theories);
- *Rejection 3:* While we cannot *deduce* values from empirical facts, we can certainly *infer* them from the facts, by various sorts of plausible reasoning, such an "inference to the best explanation" (values-as-fact theories);
- *Rejection 4:* Knowledge about values is impossible (positivism, value scepticism).

Such an analysis brings out a significant interrelationship that obtains in the theory of value between the issue of *observation* (as per (2)-rejection) and the issue of *confirmation* (as per (3)-rejection). It makes strange bedfellows.

Again, consider the apory:
1. A (cognitively) meaningful statement must be verifiable-in-principle;
2. Claims regarding what obtains in all times and places are not verifiable-in-principle;
3. Laws of nature characterize processes that obtain in all times and places;
4. Statements that formulate laws of nature are cognitively meaningful

As ever, various exists from inconsistency are available here, specifically the following four:

- *Rejection 1:* Maintain a purely semantical theory of meaning that decouples meaningfulness from epistemic considerations;
- *Rejection 2:* Accept a latitudinarian theory of verification that countenances remote inductions as modes of verification;
- *Rejection 3:* Adopt a view of laws that sees them as local regularities;
- *Rejection 4:* Maintain a radical scepticism with respect to claims regarding laws of nature, one which sees all such law-claims as meaningless.

This apory locks four very different issues into mutual relevancy: (i) the theory-of-meaning doctrine that revolves about (1), (ii) the metaphysical view regarding laws of nature at issue in (2), (iii) a philosophy-of-science doctrine regarding the nature of natural laws as operative in (3), and finally (iv) a language-oriented position regarding the meaningfulness of law claims.

An apory thus delineates a definite range of interrelated positions. It maps out a small sector of the possibility space of philosophical deliberation. And this typifies the situation in philosophical problem-solving, where, almost invariably, several distinct and discordant resolutions to a given issue or problem are available, none of which our cognitive data can exclude in an altogether decisive way. And such a situation is pervasive. Philosophical arguments can standardly be transmuted into aporetic clusters and analyzed in this light. Apories are not only frequent in philosophy but typical of the contexts in which the problems of the field arise.

This circumstance is typical of aporetic situations. Any resolution of an apory calls for the rejection of some contentions for the sake of maintaining others. Strict logic alone dictates only that something must be abandoned; it does not indicate what. No particular resolutions are imposed by abstract rationality alone—by the mere "logic of the situation." (In philosophical argumentation one person's modus ponens is another's *modus tollens*.) It is always a matter of trade-offs, of negotiation, of giving up a bit of this in order to retain a bit of that.

In all such cases, a necessity for choice is forced by the logic of the situation, but no one particular outcome is rationally con strained for us by any considerations of abstract rationality. There are forced choices but no forced resolutions. Whenever we are confronted with an aporetic cluster, a plurality of resolutions is always available. The contradiction that arises from overcommitment may be resolved by abandoning any of several contentions, so that alternative ways of averting inconsistency can always be found.

Again, consider the apory:
1. All human acts are causally determined;
2. People can and do make free acts of choice;

3. A genuinely free act cannot be causally determined (for if it is so determined, then the act is not free by virtue of this very fact)

These three theses represent an inconsistent triad to which consistency may be restored by any of three distinct approaches:
- *Deny 1:* "Voluntarism"—the exemption of free acts of the will from any causal determination (Descartes);
- *Deny 2:* "Determinism" of the will by causal constraints, dismissing free will as an illusion (Spinoza);
- *Deny 3:* "Compatibilism" of free action and causal determination—for example, via a theory that distinguishes between inner and outer causal determination and sees the former sort of determination as compatible with freedom (Leibniz).

As such examples show, a forced choice is typically upon us, seeing that any particular resolution of an aporetic cluster is bound to be simply one way among others. And the single most crucial fact about an aporetic cluster is that there will always be a variety of distinct ways of averting the inconsistency into which it plunges us. We are not only forced to choose but also constrained to operate within a narrowly circumscribed range of choice.

What If Situations

"What if" hypothesizing is a *functional* proceeding one made for a purpose. And many possibilities are at issue here:
- *Consequence-exploration.* Answering questions of the format: What would have to be the case if we stipulated that – - -. This in turn leads to two others.
- *Indirect proof* and *ad absurdum refutation.* Establishing that some thesis must be the case because the assumption of its negation would entail a contradiction or a vitiating infinite regress.
- *Dialectical refutation.* Indicating that some contention should be rejected (abandoned) because of untenable consequences.
- *Explanatory harmonization.* Showing that some contention should be accepted because otherwise some acceptable facts would remain un- or under-explained. And this in turn encompasses
- *Thought experimentation* for investigation or exploration (especially in context of scientific understanding).

The sole cogent basis for rejecting inherently meaningful assumptions as illicit and inappropriate roots in their functionality: nothing is wrong with assumptions recognized and acknowledged as false when such a supposition serves the interests of understanding.

"What if" questions are notably common in historical speculations. A plausible example of this is the counterfactual:

> If Wellington had lost at Waterloo, then Napoleon would not have been forced into (immediate) exile in St. Helena.

This admits of the following justifactory analysis:
Accepted background facts:
1. Wellington did not lose at Waterloo;
2. Wellington and Napoleon were opposed commanders at Waterloo;
3. Napoleon went into (immediate) exile;
4. Victorious commanders are not forced into immediate exile.

Assumption: not-(1): Wellington lost at Waterloo.

In attempting to make the deletions needed to restore consistency in such cases, the issue-formative assumptions are or course sacred. But even if we drop (1) in the wake of the assumption we still have a contradiction. For this assumption together with (2) entails "Napoleon won at Waterloo," so that "Napoleon was a victorious commander." And this together with (4) yields not-(3).

Given that (2) is not an issue here, our assumption of not-(1) in effect forces a choice between (3) and (4). And when we prioritize general relationships over particular facts—as is standard in counterfactual situations—we will retain (4) so that (3) must be sacrificed. And on this basis that initial counterfactual becomes validated.

Yet another plausible example of an informative historical counterfactual runs as follows:

> If Julius Caesar had not crossed the Rubican in revolt against the Roman republic, this republic would have endured far longer.

Here we have:
Accepted background facts:
1. Caesar crossed the Rubican in revolt against the Roman republic;
2. Caesar's revolt destabilized the republic and rendered it unstable and untenable;

3. Without this modality and conditionality the republic would have endured far longer.

Assume not-(1). Then by (2) Caesar's actions has not (would not have) rendered the republic unstable and untenable so that by (3) it would have endured longer. Here the conditional at issue follows simply from the hypothesis plus the limited register of salient facts.

Retention Priority

Restoring consistency in the case of counterfactuals requires a priority ranking— a "prevailing order" of retainability that governs the process of breaking the chain of inconsistency at its weakest link. It is geared to the principle of fostering informativeness via the sequential prioritization of:
- issue-definitive hypotheses, suppositions, and assumptions;
- definitive concept-conceptions and meaning-correlative relationships;
- generalizations, "laws," well-established general rules, well-confirmed causal and factual relationships;
- established trends, common relations among events, connections among particular eventuations;
- specific, particular, contingent facts and concrete eventuations.

The crux here is ow firmly and securely various claims are embedded in the overall framework of putative information.

Thus the priority situation in these *speculative* cases is exactly the reverse of the order of evidential security that obtains in the factual setting of *inductive* reasonings, where theory must give way to facts and more far-reaching theories to those that are more particular in their bearing. And in this regard specifically historical counterfactuals are in the same boat as others in that they pivot on considerations of rational economy via a Principle of the Conservation of Information that insists on optimizing our understanding of things in the face of incongruities.

Even patently false suppositions can thus be tenable in matters of counterfactual reasoning. Consider a simple example. We take ourselves to know that:
1. Bizet was French;
2. Verdi was Italian;
3. Compatriots belong to the same country. And now against this background let us make the supposition;
4. [Assume that] Bizet and Verdi were compatriots.

Obviously we now have a logical inconsistency on our hands. What to do? In the very logic of things, something has to give way: the chain of inconsistency must be broken. But where, when, and how?

To restore consistency we must clearly abandon one of (1)–(2). The other premises are safe. In arising at this position the following priorities are in place:
- *Thesis 4* is secure as the postulated assumption at issue;
- *Thesis 3* is secure as a terminological definition;
- *Thesis 3*, albeit accepted truths, have comparatively lower priority then the preceding.

Accordingly one of (1) or (2) must be abandoned, but both have exactly the same claims to retention, we are led to the conditional:

> If Bizet and Verdi had been compatriots, then either Bizet would have been Italian or Verdi would have been French.

As this shows, a cogent line of reasoning leading to a clearly acceptable conclusion can result from a patently false and contradiction-generating supposition, by suitable principles of cognitive damage control. Instructive and informative cognitive work can be accomplished under the auspices of "what if" reasoning. For in the end, the task of rational deliberations and inquiry is to extract the maximal amount of useful information from the data at our disposal—to answer the questions we have in the most extensive yet reliable way that is possible.

Distinctions to the Rescue

When an aporetic thesis is rejected, the usual course among philosophers is not to abandon it altogether, but rather to introduce a distinction by whose aid it may be retained *in part*.

Consider the following aporetic cluster, which sets the stage for the traditional "problem of evil":
1. The world was created by God;
2. The world contains evil;
3. A creator is responsible for all defects of his creation;
4. God is not responsible for the evils of this world.

On this basis, we have it that God, who is responsible for all aspects of nature, by (1), is also responsible for evil, by (3). And this contradicts contention (4). Suppose, however, that one introduces the distinction between *causal* responsibility and *moral* responsibility, holding that the causal responsibility of an agent does not necessarily entail a moral responsibility for the consequences of his acts. Then for *causal* responsibility, (3) is true but (4) false. And for *moral* responsibility, the reverse holds: (4) is true but (3) false. Once the distinction at issue is introduced, then no matter which way one turns in construing "responsibility," the inconsistency operative in the apory at issue is averted.

Thus someone who adopts this distinction can retain *all* the aporetic theses —(1) and (2) unproblematically and, as it were, half of each of (3) and (4)—each in the sense of one side of the distinction at issue. The distinction enables us to make peace in the aporetic family at issue, by splitting certain aporetic theses into acceptable and unacceptable parts.

To be sure, distinctions are not needed if *all* that concerns us is averting inconsistency; simple thesis abandonment, mere refusal to assert, will suffice for that end. But distinctions are necessary if we are to maintain informative positions and provide answers to our questions. We can guard against inconsistency by avoiding commitment. But such sceptical refrainings leave us empty handed. Distinctions are the instruments we use in the (potentially never-ending) work of rescuing our assertoric commitments from inconsistency while yet salvaging what we can.

Accordingly, one generally does not respond to cogent counterarguments in philosophy by *abandoning* one's position but rather by making it more sophisticated—by *complicating* it. One can never entrap any philosophical doctrine in a finally and decisively destructive inconsistency, because a sufficiently clever exponent can always escape from difficulty by means of suitable distinctions.

Consider, for example, the following aporetic cluster:
1. Only something real can produce real effects (only real causes are really causes);
2. Delusions and illusions can produce real effects;
3. Delusions and illusions are not real.

One easy way out is to reject (3), but to do so conservatively via a distinction that insists that delusions and illusions are indeed real as such (viz., as subjective mental episodes of a certain sort)?it is just that their objects are not real. So if Jones recoils in horror before an imaginary snake, the subjective side of this process (the illusory snake-imagining) is real enough and so acts as a real cause; it's just that the (ex hypothesi nonexistent) snake is—being nonexistent —impotent to produce any effect. The snake itself does not really exist. It is only a figment of the imagination—though as such (viz., as an illusory idea) it is indeed real and can thus produce real effects, even as (1) would have it. This is clearly an effective way of escaping the aporetic inconsistency at issue.

Consider the apory detailed in Display 5.1 below. Clearly, this situation confronts us with an outright contradiction. Something must give way. But what? Since (1), (2) and (6) are merely hypotheses here, they stand secure. Three alternatives are thus open: (i) to abandon (3), (ii) to abandon (8), and (iii) to abandon (5) as untenable in the presence of (1) and (2). Clearly, however, none of these options is particularly attractive. We do not want to abandon (3) and dissociate belief from action in the context of rationality.

Display 5.1: A Paradox of Rational Belief

---

1. One believes it to be the case that p
2. One is a rational agent
3. Rational agents act on their beliefs
4. One will act (in any and all circumstances) on one's belief that *p* (from (1)-(3))
5. One recognizes (concedes, allows) that there is some small chance that *p* might be false
6. One' is offered a bet that will pay one cent if *p* is true and invoke an awful catastrophe (say the end of organic life in the universe) if *not-p* is true
7. One would bet on *p* in this case (from (4), (6))
8. Rational agents are (somewhat) Bayesian. They guide their actions by the balance of risk and return. And so they do not accept minute inducements to run even small risks of (sufficiently great) disasters
9. One would bet on *not-p* in this case (from (5), (6), (8))

Again, we do not want to abandon (8). We are rightly reluctant to forego the Bayesian approach to rational decision making. Nor is it all that easy to abandon (5). For it is not an appealing prospect to hold that rational believers cannot concede any prospect or possibility that their beliefs might be false and are constrained to see all their beliefs as absolutely and definitively certain. How can we exit from difficulty?

As is so often the case with such theoretical difficulties, the way out lies through the door of a distinction. Our beliefs are not all of a piece. There are some things we believe-to-be-absolutely-certain (C-beliefs). We view these as totally secure and utterly safe. We would bet literally everything on them. With respect to these beliefs—but only these the inference from (1)–(3) to (4) holds good. But such wholly unconditional beliefs, of course, are few and far between. The rational man uses due epistemic caution. Most of what we believe we believe-to-be-plausible (P-beliefs). We view these as adequately secure and relatively safe only against all realistic possibilities that something might go wrong, but not against all conceivable possibilities. We would obviously not put everything at risk for these P-beliefs.

On the basis of this distinction, then, we are able to avert the paradox of Table I. The reading of "belief" in (1) that authorizes the move to (4) via (3) presumes that we are dealing with C-beliefs. But the reading of "belief' in (1) that allows us to invoke (5) is predicated on its being P-beliefs that are at issue. The distinction between C-beliefs and P-beliefs averts the paradox at issue. It enables us to effect a reduction of that over-rich family of (inconsistent) initial. Commitments. In this way it typifies the role of distinctions in philosophical deliberations.

More on Distinctions

The history of philosophy is replete with distinctions introduced to avert the aporetic difficulties inherent in oversimplification. Already in the dialogues of Plato, the first systematic writings in philosophy, we encounter distinctions at every turn. In Book I of Plato's *Republic*, for example, Socrates' interlocutor quickly falls into the following apory:
1. Rational people always pursue their own interests;
2. Nothing that is in a person's interest can be disadvantageous to him;
3. Even rational people sometimes do things that prove disadvantageous.

However, the evident inconsistency that arises here can be averted by distinguishing between two senses of the "interests" of a person—namely the real and the apparent, what is *actually* advantageous to him and what he merely *thinks* to be so. Again, in the discussion of "nonbeing" in Plato's *Sophist*, the Eleatic stranger entraps Socrates in an inconsistency from which he endeavors to extricate himself by distinguishing between "nonbeing" in the sense of not existing *at all* and in the sense of not existing *in a certain mode*, that is, between absolute and sorted nonexistence. Throughout the Platonic dialogues present a dramatic unfolding of one distinction after another.

Distinctions enable us to implement the idea that a satisfactory resolution of problems of aporetics inconsistency must somehow make room for all parties to the contradiction. The introduction of distinctions thus represents a Hegelian ascent-rising above the level of antagonistic positions to that of a "higher" conception, in which the opposites are reconciled. In introducing the qualifying distinction, we abandon the initial thesis and move toward its counterthesis, but we do so only by way of a duly hedged synthesis. In this regard, distinction is a "dialectical" process.

This role of distinctions is also connected with the principle that is sometimes designated as "Ramsey's Maxim." With regard to disputes about fundamental questions that do not seem capable of a decisive settlement, Frank Plumpton Ramsey wrote: "In such cases it is a heuristic maxim that the truth lies not in one of the two disputed views but in some third possibility which has not yet been thought of, which we can only discover by rejecting something assumed as obvious by both disputants."[14] On this view, too, distinctions provide for a higher synthesis of opposing views. They prevent thesis abandonment from being an entirely negative process, affording us a way of salvaging something, of "giving credit where credit is due" even to those contentions we ultimately reject. They make it possible to remove inconsistency not just by the brute force of thesis rejection, but by the more subtle and constructive device of thesis qualification.

A distinction reflects a *concession,* an acknowledgement of some element of acceptability in the thesis that is being rejected. However, distinctions always involve us in bringing a new concept onto the stage of consideration and thus put a new topic on the agenda. They accordingly always afford invitations to carry the discussion further, opening up new issues that were heretofore conceptually inaccessible. Distinctions are the doors through which philosophy passes into new topics and problems. New concepts and new theses come to the fore.

Philosophical distinctions are thus creative innovations devised to ensure the acceptability of our claims. There is nothing routine or automatic about them—their discernment is an act of inventive ingenuity. They do not elaborate preexisting ideas but introduce new ones. They not only provide a basis for understanding better something heretofore grasped imperfectly, but they shift the discussion to a new level of sophistication and complexity. Thus, to some extent they "change the subject." (In this regard they are like the conceptual innovations of science, which revise rather than explain prior ideas.)

The continual introduction of new concepts via new distinctions means that the ground of philosophy is always shifting beneath our feet. New distinctions for our concepts and new contexts for our theses alter the very substance of the old theses. The development is dialectical—an exchange of objection and response that constantly moves the discussion onto new ground. The resolution of antinomies through new distinctions is a matter of innovations whose outcomes cannot be foreseen.

Distinctions avert conflict and yet they seldom settle controverted issues in a definitive way. For they always leave a crucial evaluative issue hanging in the air: the issue of *priority.* The pivotal question always arises, given that the term T can be split apart into the two senses T_1 and T_2, which of these two captures the

---

[14] Ramsey 1931, pp. 115–16.

"standard" or "normal" use of the word? Which construction is it that we should generally give to the equivocal word when we meet it in the relevant discussions? (For example: is it belief-as-true or belief-as-plausible that is at issue in standard cases?) Which sense *predominates?*

Consider the following apory:

1. Only observationally verifiable sentences are (genuinely) meaningful. (Positivism);
2. The speculative claims of traditional metaphysics are not observationally verifiable;
3. The speculative claims of traditional metaphysics are meaningful. (Metaphysical traditionalism)

Given that (2) is "fact of life," we are driven to a choice between (1) and (3). Now a peacemaker might propose a distinction here, offering the following proposal:

> Let us introduce the (somewhat technical) idea of *empirical* meaningfulness—that is, let us distinguish between what is specifically empirically meaningful ('experientially resolvable' in some way) and what is not. Then one can accept (1) and abandon (3) in this particular sense, while, on the other hand, retaining (3) and abandoning (1) with respect to 'loose, old-fashioned meaningfulness-at-large.'

But it is clear that such a distinction, which enables us to "have it both ways," will not really make peace between the metaphysical traditionalist and his positivist adversary. Even while agreeing to "split the difference" in the face of the distinction, the positivist will say in his heart: "It is *empirical* meaningfulness that really counts, it is in this that true-blue authentic meaningfulness consists." The metaphysician, on the other hand, will say: "This idea of '*empirical* meaningfulness' is a mere technical construction that is really beside the point. It is meaningfulness-at-large that captures the authentic core of the idea." There is now a fight, of sorts, for the right to the succession. Each of the new distinction-generated conceptions seeks to establish itself as the principal heir of the root concept. The quarrel now becomes one of which side represents the prime, main, most important aspect of the root distinction-antecedent idea?

Consistency Maintenance Precedence and Priority

Apories engender forced choices—choices which distinctions can mitigate but can never wholly avert. For what we need to do in order to effect a reasoned choice among the alternatives is to establish some *priority* or *precedence* among the data: to implement the idea that while they are all "acceptable" in a credibility-oriented sense, some are *more* acceptable than others. In situations of potential conflict, we must recognize some have lesser claims on us for retention than others. Interesting ramifications lurk here.

In philosophy, the evidentially factual, purely imposed by more consistency constraints rational invariably underdetermine the resolution of our problems. To be sure, the cognitive-evidential situation is such that considerations of abstract rationality require us to make choices ("forced choices"). But there are no *forced resolutions,* for we are never constrained to a *particular* mode of inconsistency elimination in the cognitive situation at hand, at least not by considerations of abstract rationality alone. Concrete resolutions are always *underdetermined* by considerations of *cognitive* rationality; they become determined only when considerations of *evaluative* rationality come upon the scene. Philosophical problem solving is, in the final analysis, an evaluative matter—though, to be sure, it is not aesthetic or ethical values that are at issue but specifically cognitive values that relate to matters of importance, centrality, significance, or the like. In philosophy, our problem-resolutions always involve us in issues of precedence and priority. Cost-benefit parameters like "plausible," "natural," come into prominence via such cognitive values as simplicity, economy, uniformity, harmoniousness, and the like.

The issue that we confront in apory-resolution is thus one of priority—and ultimately one of *evaluation.* What it takes to resolve an apory is a matter of setting priorities—of substantiating a preferential choice, albeit one made on the basis of evaluative factors that relate to specifically informative matters. Probative values serve as the decisive factors here.

The resolution of apories is an exercise in cost-benefit analysis. No matter which way we turn in removing inconsistency, we pay certain costs in terms of thesis abandonment and concept-complexification to achieve the benefits of retaining various aspects of the *status quo ante.* We pay the cost of complexification for the benefit of continuing our commitment to what seems "only plausible and natural" from a less reflective standpoint. Some appeal of a fundamentally evaluative nature is always ultimately involved. This ultimately, is why philosophical disputes are so recalcitrant.

Coda

Some may see such a position as having ominously sceptical implications for the status of philosophy—as undermining the validity of the whole enterprise. But this view will itself be based on a very questionable evaluative position. For the present deliberations regarding evidential underdetermination only indicate the indecisiveness of philosophical deliberations for someone who thinks that only resolutions founded on strictly evidential considerations are really worth having—that problem-resolutions involving the invocation of cognitive values are somehow inferior, questionable, and not really worthwhile. Curiously enough, such disparagement of the evaluative domain itself represents a thoroughly evaluative posture—and a highly problematic one at that.[15]

---

[15] The essay is a much-revised version of a paper originally published in *The Review of Metaphysics*, vol. 41 (1989), pp. 253–297.

6 Metaphilosophical Coherentism

Introduction

Metaphilosophical deliberation usually inclines towards a view of philosophy itself that locates the goal of this enterprise in the intellectual construction of a cogent and comprehensive account of the nature and grounding human mind's experience of its world. The coherentism to which idealism in general inclines is operative here as well. However, such a view of philosophizing's mission soon constrains the project to confront the implications of the complexity of human experience—its immense diversity and variability. Conflicting tendencies and tensions are present throughout. Any attempt at the rational systematization of experience is bound to confront its inner stresses and strains that brings an aporetic complexity to the fore. Against the background of this general line of thought, there is a good case for three interconnected theses regarding philosophical methodology:
1. Problem-solving in philosophical inquiry standardly involves the resolution of aporetic conflicts among individually plausible but collectively inconsistent propositions; and
2. The resolution of such conflicts calls for fundamentally evaluative mechanisms of precedence and priority—issues which here cannot be resolved by purely evidential means, but require evidence-transcending evaluative resources; yet nevertheless
3. Philosophy is a rational enterprise that does not degenerate into an indifferentist relativism of rationally arbitrary individual preferences.[16]

Critical responses to this position have taken two principal forms: Some have maintained that given a commitment to the evaluativeness of (1) and (2) one is not entitled to the rationalism of (3), so that my position would ultimately be a crypto-relativism.[17] Others have argued that given a commitment to the rationalism of (1) and (3) one is not entitled to the evaluativeness of (2), so that

---

Note: This essay is a revision of a paper of the same title originally published in *Idealistic Studies*, vol. 27 (1997), pp. 131–141.

16 Regarding this position, see Rescher 1985 and 1994.
17 See Wallace 1992, pp. 222–23.

I should, strictly taken, be committed to an evidentialist absolutism that dispenses with evidence-transcending evaluations.[18]

The present discussion proposes to maintain that those there aforementioned theses are indeed compatible, and that the evaluative commitment of thesis (2) can indeed be squared with thesis (3)'s rejection of rationalistic relativism.

Apory Resolution as Matters of a Precedence and Priority

Consider, for the sake of an historical example, the following group of contentions, all of which were viewed favorably by Presocratic philosophers, bring part of the received data of the time:
1. Reality is one: real existence is homogenous;
2. Matter is real (self-subsistent);
3. Form is real (self-subsistent);
4. Matter and form are distinct (heterogeneous).

Here (2) – (4) insist that reality is heterogeneous, thereby contradicting (1). The whole of the group (1) – (4) accordingly represents an aporetic cluster that reflects a cognitive overcommitment. And this situation is typical: the problem context of philosophical issues standardly arises from a clash among individually tempting but collectively incompatible overcommitments. Philosophical issues standardly center about an aporetic cluster of this sort—a family of plausible theses that is assertorically *overdeterminative* in claiming so much as to lead into inconsistency.

In such cases, something obviously has to go. Whatever favorable disposition there may be toward these plausible theses, they cannot be maintained in the aggregate. We are confronted by a (many-sided) cognitive dilemma and must find one way out or another. In particular, we can proceed:
- To reason from (2)-(4) to the denial of (1);
- To reason from (1), (3), (4) to the denial of (2);
- To reason from (1), (2), (4) to the denial of (3);
- To reason from (1)-(3) to the denial of (4).

An apory gives rise to a group of valid arguments leading to mutually contradictory conclusions, yet each having only plausible theses as premises. It is clear in

---

18 Haak 1987.

such cases *that* something has gone amiss, though it may well be quite unclear just where the source of difficulty lies.[19]

The resolution of such an aporetic situation obviously calls for abandoning one (or more) of the theses that generate the contradiction. Unexceptionable as these theses may seem, one or another of them has to be jettisoned. The restoration of consistency is an imperative. And the problem is that there are always alternative ways of doing this. Thus, the ancient Greek philosophers confronted the following range of possibilities:

- *Denial 1:* Pluralism (Anaxagoras) or form/matter dualism (Aristotle);
- *Denial 2:* Idealism (the Eleatics, Plato);
- *Denial 3:* Materialism (atomism);
- *Denial 4:* Dual-aspect theory (Pythagoreanism).

We are plunged into such dilemmas by cognitive overcommitment. Too many jostling contentions strive for our approbation and acceptance. And this state of affairs is standard in philosophy—indeed the standard impetus to philosophical reflection. The task of philosophy, as Socrates clearly saw, is to work our way out of the thicket of inconsistency in which we are entangled by our presystemic beliefs.

The requirements for conflict-resolution lie at the forefront. With philosophical apories some plausible-seeming thesis must be made to give way to others. A process for deciding issues of precedence and priority—for a rationally cogent basis for the cooperative assessment of claims. Conflict resolution in the face of aporetic inconsistency is a matter of evaluation, of making appraisals in point of precedence and priority in a way that enables us to decide which one among conflicting plausible contentions must give way to the others. Without such an evaluative mechanism we are left in a condition of cognitive vacuity.

It is clear, however, that the evaluation at issue most be cognitive rather than affective. It is a matter not of what we *like* or what we would ask for—but rather one of realizing which among conflicting alternatives is best spoken for by such factors of rational support as we can come by.

---

19 This linkage of a philosophical thesis to an aporetic cluster in which it stands in correlative apposition to its rivals makes it plausible to hold the paradoxical-sounding view that "the argument for [and against] a philosophical statement is always a part of its meaning" (Johnstone 1959, p. 32). For the position at issue only comes to be defined as such in the context of the counterpositions it proposes to exclude; in philosophy, Spinoza's dictum holds good: *ominis determinatio est negatio*.

The Thrust of Coherentism

Can we make our forced cognitive choices on strictly *evidential* grounds? Can straightforward evidentiation as we know it in other contexts in our everyday life and our science do the job in philosophy as well. Unfortunately not. In philosophical contexts, a cognitive prioritization of potentially conflicting theses cannot be accomplished by purely evidential means but requires other philosophical resources. The reason for this lies ultimately in the nature of the subject itself. For unlike biology or physics, philosophy is a reflexive subject: questions *about* the discipline fall within the discipline itself. Meta-philosophy is a constituent part of philosophy. And, specifically, *how evidentiation itself works in philosophical contexts* is an issue that is at once itself evidence-transcending (i.e., beyond the prospect of settlement on strictly evidential grounds) and clearly philosophical in its nature and its ramifications. Moreover, philosophical apories arise in a context where the evidence has already spoken and has done pretty well all it can do. And so, while we know (thanks to the inconsistency) *that* something is wrong, we cannot deploy the available evidence to determine *what* has gone amiss.

It follows that the evaluative priority-determination at issue in the resolution of philosophical apories cannot be realized by purely evidential means but calls for the availability of other, evidence-transcending resources. This inconvenient-seeming fact seems is part of the reality with which philosophy has to come to terms.

But if *evidence* and such does not accomplish philosophy's priority-establishing work, then what does?

The answer is *coherence*. Smoothness of fit within the overall fabric of our experience. Philosophical substantiation is not a matter of finding the best evidentiated conclusion (which makes sense only where the evidence is well-grounded and consistent) but rather one of finding what is in line with the smoothest overall systematization of "the data" (which, as we know, is namely tentative and collectively inconsistent). It is a matter not of *evidentiation*, of inference to the best confirmed, but rather one but of "inference to the best systematization." (Exactly this sort of rationale explains why philosophical idealists have generally been attached to a coherence criteriology of truth.)

Given inputs of this sort—plausible data—the coherence analysis sets out to sift through these truth-candidates with a view to minimizing the conflicts that may arise. Its basic mechanism is that of best-fit considerations, which brings us to a coherentism whose stance is essentially as follows:

That family embracing the truth candidates which are maximally attuned to one another is to count—on this criterion of over-all mutual accommodation—as best qualified for acceptance as presumably true, implementing the idea of compatibility screening on the basis of "best-fit" considerations. Mutual coherence becomes the arbiter of acceptability which make the less plausible alternatives give way to those of greater plausibility. The acceptability-determining mechanism at issue proceeds on the principle of optimizing our admission of the claims implicit in the data, striving to maximize our retention of the data subject to the plausibilities of the situation.[20]

Against this background, the general strategy of philosophical coherentism lies in three-step procedure:
1. To gather in all of the "data" (in the present technical sense of this term);
2. To lay out all the available conflict-resolving options that represent the alternative possibilities that are cognitively at hand;
3. To choose among these alternatives by using the guidance of plausibility considerations, invoking (in our present context) the various parameters of systematicity as indices of plausibility.

In this way, the coherence theory implements F. H. Bradley's dictum that *system* (i.e., systematicity) provides a test-criterion most appropriately fitted to serve as arbiter of truth. In philosophical argumentation it is systematization of question-resolving conjecture *with the data* that provides the best available standard of cogency.

Plausibility as a Guide

The concept of plausibility affords our prime guide in philosophical problem-solving. Here plausibility and presumption are closely interrelated in rational argumentation via the principle:

> Presumption favors the most *plausible* of rival alternatives—when indeed there is one. This alternative will always stand until set aside (by the entry of another, yet more plausible, presumption).

The operation of this rule creates a key role for plausibility in the theory of reasoning and argumentation. In the fact of discordant considerations, one "plays safe" in one's cognitive involvements be endeavoring to maximize the plausibility levels achievable in the circumstances. Such an epistemic policy is closely

---

20 The formal mechanism of best-fit analysis is described more fully in Rescher 1973 1976.

analogous to the *prudential* principle of action—that of opting for the available alternative from which the least possible harm can result. Plausibility-tropism is an instrument of epistemic prudence.

The plausibility of a thesis will not be a measure of its *probability*—of how likely we deem it, or how surprised we would be to find it falsified. Rather, it reflects the prospects of its being fitted into our cognitive scheme of things in view of the standing of the source or principles that vouch for its inclusion herein. The core of the present conception of plausibility is the notion of the extent of our cognitive inclination towards a proposition—of *the extent of its epistemic hold upon us* in the light of the credentials represented by the bases of its credibility. The key issue is that of how readily the thesis in view could make its peace within the overall framework of our cognitive commitments.

The standing of cognitive sources in point of their authoritativeness affords one major entry point to plausibility. In this approach, a thesis is more or less plausible depending on the reliability of the sources that vouch for it—their entitlement to qualify as well-informed or otherwise in a position to make good claims to credibility.[21] However, the plausibility of contentions is often based not on a thesis-warranting *source* but a thesis-warranting *principle*. Here inductive considerations may come prominently into play; in particular such warranting principles are the standard inductive desiderata: simplicity, uniformity, specificity, definiteness, determinativeness, "naturalness," etc. On such an approach one would say that the more simple, the more uniform, the more specific a thesis—either internally, of itself, or externally, in relation to some stipulated basis—the more emphatically this thesis is to count as plausible.

For example, the concept of *simplicity* affords a crucial entry point for plausibility considerations. The injunction "Other things being anything like equal, give precedence to simpler hypotheses vis-a-vis more complex ones" can reasonably be espoused as a procedural, regulative principle of presumption, rather a metaphysical claim as to "the simplicity of nature." On such an approach, we espouse not the Scholastic adage "Simplicity is the sign of truth" (*simplex sigilium veri*), but its cousin, the precept "Simplicity is the sign of plausibility" (*simplex sigilium plausibili*). In adopting this policy we shift the discussion from the plane of the constitutive/descriptive/ontological to that of the regulative/methodological/prescriptive.

---

[21] This view of plausibility in terms of general acceptance either by the *consensus gentium* or by the experts ("the wise") was prominent in Aristotle's construction of the plausible (*endoxa*) in Book I of the *Topics*.

Again, uniformity can also serve as a plausibilistic guide to reasoning. Thus consider the *Uniformity Principle:*

> In the absence of explicit counterindications, a thesis about unscrutinized cases which conforms to a patterned uniformity obtaining among the data at our disposal with respect to scrutinized cases—a uniformity that is in fact present throughout these data—is more plausible than any of its regularity-discordant contraries. Moreover, the more extensive this pattern-conformity, the more highly plausible the thesis.

This principle is tantamount to the thesis that when the initially given evidence exhibits a marked logical pattern, then pattern-concordant claims relative to this evidence are—*ceteris paribus*—to be evaluated as more plausible than pattern-discordant ones (and the more comprehensively pattern-accordant, the more highly plausible). This rule implements the guiding idea of the familiar practice of judging the plausibility of theories of theses on the basis of a "sufficiently close analogy" with other cases.[22] (The uniformity principle thus forgoes a special rule for normality—reference to "the usual course of things"—in plausibility assessment.)[23]

In general, then, the closer its fit and the smoother its consonance with our cognitive commitments, the more highly plausible the thesis.[24] Ordinarily, the removal of a highly plausible thesis from the framework of cognitive commitments would cause a virtual earthquake; removal of a highly implausible one would cause scarcely a tremor; in between we have to do with varying degrees of readjustment and realignment.

The coherentist criterion of accent ability accordingly assumes an entirely *inward* orientation; it does not seek to compare the truth candidates directly with "the facts" obtaining outside the epistemic context; rather, having already gathered in as much information (and this will include also *misinformation*) about

---

22 All this, of course, does not deal with questions of the status of this rule itself and of the nature of its own justification. It is important in the present context to stress the *regulative* role of plausibilistic considerations. This now becomes a matter of *epistemic policy* ("Give priority to contentions which treat like cases alike") and not a metaphysical laden contention regarding the ontology of nature (as with the—blatantly false—descriptive claim "Nature is uniform"). The plausibilistic theory of inductive reasoning sees uniformity as a *regulative principle of epistemic policy* in grounding our choices, not as a *constitutive principle* of ontology. As a "regulative principle of epistemic policy" its status is *methodological*—and thus its justification is in the final analysis pragmatic. See the Rescher 1976.
23 Gonseth 1947. The work of Herbert A. Simon is an important development in this area: see Simon 1966. On the larger ramifications of the issue see the Rescher 1994.
24 For a closer study of the notion of plausibility and its function in rational argumentation, see Rescher 1976.

the facts as possible, it seeks to sift the true from the false *within* this body. On this approach, the validation of an item of knowledge—the rationalization if its inclusion alongside others within "the body of our knowledge"—proceeds by way of exhibiting its interrelationships with the rest: they must all be linked together in a connected, mutually supportive way (rather than having the form of an inferential structure built up upon a footing of rock-bottom axioms). On the coherentist's theory, justification is not a matter of derivation but one of systematization. We operate, in effect, with the question: "justified" = "systematized." The coherence approach can be thought of as representing, in effect, the systems-analysis approach to the criteriology of truth. And idealists generally view this modus operandi as our best resource in philosophy.

The Empiricism Aspect

Data—construed along the lines of the preceding discussion—play a pivotal role in the coherentist criteriology of truth via optimal systematization. The entire drama of a coherence analysis is played out within the sphere of propositions, in terms of the sorts of relationships they have to one another. Now there is—in any event—enough merit in a correspondence account of truth that an appropriate consonance must obtain between "the actual facts of the matter" and a proposition regarding them that can qualify as true to the criticism "why should mere coherence imply truth?" one can and should reply: what is at issue here is not *mere* coherence, but coherence *with the data*. It is not with bare coherence as such (whatever that would be) but with data-directed coherence that a truth-making capacity enters upon the scene. But, of course, datahood only provides the building blocks for truth-determination and not the structure itself. Coherence plays the essential role because it is to be through the mediation of coherence considerations that we move from truth-candidacy and presumptions of factuality to truth as such. And the procedure is fundamentally non-circular: we need make no imputations of truth at the level of data to arrive at truths through application of the criterial machinery in view.

In his exposition of the coherence theory, Brand Blanshard has written:

> Granting that propositions, to be true, must be coherent with each other, may they not be coherent without being true? ... Again, a novel, or a succession of novels such as Galsworthy's *Forsyte Saga*, may create a special world of characters and events which is at once extremely complex and internally consistent; does that make it the less fictitious? ... This objection, like so many other annihilating criticisms, would have more point if anyone had ever held the theory it demolishes. But if intended to represent the coherence theory as responsibly advocated, it is a gross misunderstanding. That theory does not hold that

> any and every system is true, no matter how abstract and limited; it holds that one system only is true, namely the system in which everything real and possible is coherently included. How one can find in this the notion that a system would still give truth if, like some arbitrary geometry, it disregarded experience completely, it is not easy to see.[25]

This key passage, intended to answer a basic objection, leaves matters in a badly muddled state. Just where is coherence to be operative? In "the system in which everything real and possible is coherently included"? But here—in this *all-inclusive* system—there is no difference drawn or to be drawn between the actually real and the merely possible with respect to coherence: this exactly is the force of the initial objection. Yet Blanshard's position is at bottom correct. The coherence theory would indeed be deficient if it held "that a system would still give truth if … is disregarded experience completely." Our recourse to data is intended to supply just this requisite of a recourse to "experience."

The concept of a datum along the general lines explicated in the preceding discussion is certainly no newcomer to epistemology. Coherence theorists and others have articulated conceptions of much this same sort, F. H. Bradley himself being a prime case in point. In his essay "On Truth and Coherence,"[26] Bradley introduces the concept of a *fact* so as to have it play a role closely akin to that of a truth-candidate in our sense. First of all, Bradleyan "facts"—let us call them *B*-facts—differ from everyone else's facts in not necessarily being factual, i.e. true. Typical, for Bradley, are the "facts of perception and memory," which need not, of course, be true at all, but are at best purportedly or presumptively veridical:

> These facts of perception [and memory], I further agree, are at least in part irrational [and so false]. … [Yet] I do not believe that we can make ourselves independent of these non-relational data. But, if I do not believe all this, does it follow that I have to accept independent facts [i.e., facts true independently of all other considerations]? Does it follow that perception and memory give one truths which I must take up and keep as they are given me, truths which in principle cannot be erroneous? This surely would be to pass from one false extreme to another. … I therefore conclude that no given fact is sacrosanct. With every fact of perception or memory a modified interpretation is in principle possible, and no such fact therefore is free from all possibility of error.[27]

Bradley espouses—with respect to the limited range of perception, memory, and the testimony of others—a notion of "fact" according to which the facts do not

---

25 Blanshard 1939, pp. 275–6.
26 Bradley 1914, pp. 202–18.
27 Bradley 1914, pp. 203–4. See also the discussion of Bradley's position in Ewing 1934, especially pp. 239–40.

automatically qualify as truths at all but at best as possible or potential truths.[28] Despite its limitations in scope, Bradley's conception of a "fact" is clearly a precursor of—indeed almost a paradigm for—our own conception of a datum.

One acute critic has made the following charge against traditional formulations of the coherence theory:

> It is on this point particularly that the historical coherence theory appears to be ambiguous; it seems never possible to be sure, in presentations of that conception, whether 'coherence' implies some essential relation to *experience*, or whether it requires only some purely logical relationship of the statements in question. Indeed, the so-called 'modern logic', associated with this theory, is such as totally to obscure the essential distinction between analytic truths of logic and those empirical truths we can only be assured by some reference beyond logic to given data of sense.[29]

The presently contemplated version of the coherence theory is immune to this criticism. For the requisite "essential relation to *experience*" is provided by the essential reliance upon *data*—i.e. by restrictive use of only certain propositions as data. The "purely logical relationship of the statements in question" in terms of which the conception of coherence is implemented comes into play only after these data are in hand, in providing the mechanism for determining some among them to be truths.

The concept of a datum thus does a critically important job for the coherence theory of truth. It serves to provide an answer to the question "Coherent with what?" without postulating a prior category of fundamental truth. It provides the coherence theory with grist to its mill that need not itself be the product of some preliminary determinations of truth. A reliance upon data makes it possible to contemplate a coherence theory that produces truth not *ex nihilo* (which would be impossible) but yet from a basis that does not itself demand any prior determinations of truthfulness as such. A coherence criterion can, on this basis, furnish a mechanism that is *originative* of truth—that is, it yields truths as outputs without requiring that truths must also be present among the supplied inputs.

---

28 Compare C. I. Lewis's thesis that "whatever is remembered, whether as explicit recollection or merely in the form of our sense of the past, is *prima facie* credible because so remembered" (Lewis 1946, p. 334). This line of thought was developed earlier in substantial detail in A. Meinong's important essay "Zur erkenntnistheoretischen Würdigung des Gedächtnisses" in Meinong 1933. Meinong there argues that memory-derived judgements must be accorded 'immediate presumptive evidence' (p. 207). This conception of the *presumptively evident* is clearly yet another precursor of the conception of prima facie truth.
29 Lewis 1962, p. 339.

It is sometimes objected that coherence cannot be the standard of truth because there we may well arrive at a multiplicity of diverse but equally coherent structures, whereas truth is of its very nature conceived of as unique and monolithic. Bertrand Russell, for example, argues in this way:

> ... there is no reason to suppose that only *one* coherent body of beliefs is possible. It may be that, with sufficient imagination, a novelist might invent a past for the world that would perfectly fit on to what we know, and yet be quite different from the real past. In more scientific matters, it is certain that there are often two or more hypotheses which account for all the known facts on some subject, and although, in such cases, men of science endeavor to find facts which will rule out all the hypotheses except one, there is no reason why they should always succeed.[30]

One must certainly grant Russell's central point: however the idea of coherence is articulated in the abstract, there is something fundamentally undiscriminating about coherence taken by itself. Coherence may well be—nay certainly is—a descriptive feature of the domain of truths: they cohere. But there is nothing in this to prevent propositions other than truths from cohering with one another: Fiction can be made as coherent as fact: truths surely have no monopoly of coherence. Indeed "it is logically possible to have two different but equally comprehensive sets of coherent statements between which there would be, in the coherence theory, no way to decide which was the set of true statements."[31] In

---

30 Russell 1912, p. 191. Or compare M. Schlick's formulation of this point: 'Since no one dreams of holding the statements of a story book true and those of a text of physics false, the coherence view fails utterly. Something more, that is, must be added to coherence, namely, a principle in terms of which the compatibility is to be established [sc. as factual], and this would alone then be the actual criterion'. Schlick 1959b (see p. 216).

31 White 1967, pp. 130–3, esp. p. 131. One critic of the coherence theory elaborates this important point with demonstrative clarity as follows: "That in the end only one sufficiently comprehensive system of statements would be found consistent, is a suggestion which runs counter to obvious facts about the nature of consistency and of systems; probably it strikes us as plausible because we are such poor liars, and are fairly certain to become entangled in inconsistencies sooner or later, once we depart from the truth. A sufficiently magnificent liar, however, or one who was given time and patiently followed a few simple rules of logic, could eventually present us with any number of systems, as comprehensive as you please, and all of them including falsehoods. Insofar as it is possible to deal with any such notion as 'the whole of the truth', it is the Leibnizian conception of an infinite plurality of possible worlds which is justified, and not the conception of the historical coherence theory that there is just one all-comprehensive system, uniquely determined to be true by its complete consistency. ... Thus if we start with any empirical [i.e., contingent] belief or statement 'P', we shall find that one or other of every pair of further empirical statements, 'Q' and 'not-Q', 'R' and 'not-R', etc., can be conjoined with 'P' to form a self-consistent set. And exactly the same will likewise be true of its contradictory 'not-

consequence, coherence cannot of and by itself discriminate between truths and falsehoods. Coherence is thus seemingly disqualified as a means for *identifying* truths. Any viable coherence theory of truth must make good the claim that despite these patent facts considerations of coherence can—somehow—be deployed to serve as an indicator of truth. And so a sensible coherentism can arrest those purported difficulties. It would look not to coherence pure and simple for its criterion of truth, but to coherence with the data of experience. It thus renders Russell's objection effectively irrelevant.

A further criticism is developed by pressing hard upon the question "coherence with what?" Is this to mean coherence with *everything*—with *all* other propositions that can be enunciated? That is patently impossible: as a body, the totality of meaningful propositions is certainly inconsistent—and so incoherent. Is it merely coherence with *something*—i.e. with some *other* propositions—that is asked for? That clearly will not do. A novel or science fiction tale or indeed any other sort of made-up story can be perfectly coherent. To say simply that a proposition coheres with *certain* others is to say too little. *All* propositions will satisfy this condition, and so it is quite unable to tell us anything that bears upon the question of truth. It would be quite senseless to suggest that a proposition's truth resides in "its coherence" alone. Its coherence is of conceptual necessity a relative rather than an absolute characteristic. Coherence must always be coherence *with something:* the verb "to cohere with" requires an object just as much as "to be larger than" does. We do not really have a coherence theory in hand at all, until the *target domain* of coherence is specified. Once this has been done, we *may* very well find that the inherent truth-indeterminacy of abstract coherence —its potential failure to yield a unique result—has been removed. Now it must be said in their defense that the traditional coherence *per se*, but have insisted that it is specifically "coherence with our experience" that is to be the standard of truth.[32] The coherence theory of the British idealists has never abandoned altogether the empiricist tendency of the native tradition of philosophy.

P'. *Every empirical supposition, being a contingent statement, is contained in some self-consistent system which is as comprehensive as you please.* And as between the truth of any empirical belief or statement 'P' and the falsity of it (the truth of 'not-P') consistency with other possible beliefs or statements, or inclusion in comprehensive and self-consistent systems, provides no clue or basis of decision" (Lewis 1962, pp. 340–1).

32 See Ewing 1934, p. 238, as well as his later essay on "The Correspondence Theory of Truth" where he writes, "that coherence is the test of truth can only be made plausible if coherence is interpreted not as mere internal coherence but as coherence with our experience" (Ewing 1968, pp. 203–4). For an author of the earlier period see H. H. Joachim who writes: "Truth, we said, was the systematic coherence which characterized a significant whole. And we proceeded to

Coherentism and Its Alternatives

As cognitive theorists generally see it, there are three principal ways to substantiate contentions in rational inquiry: (1) by the data of observation, (2) by rational inference from other, already substantiated claims, (3) by plausible conjectures that harmonize and coordinate otherwise substantiated considerations.

Those who reject (3) while accepting only (1) and (2) as appropriate are *empiricist foundationalists*, in that they ultimately ground all acceptance in observation of some sort, be it sensory or immediate ("intuition"). Those who go beyond (2) to accept (3) and so take contextually harmonious conjecture as sufficient for warranted acceptability are known as *systemic coherentists*. For their acceptability is a matter of comprehension fitness and harmonization in the ordered manifold of our commitments.

The crux of empiricism lies in seeing knowledge as a systematization of observation.

The nearest familiar analogy for coherentism is the idea of circumstantial evidence in law. The contrast here is that between the direct evidence of personal observation and immediate experimental contact and indirect or "circumstantial" evidence that serves to provide a coherent picture of relevant events. The crux of coherentism lies in seeing knowledge as a manifold formed by the considerations of systemic coordination and contextual unity.

Truth as "Adequatio ad Rem"

Coherentism's main issue is the so-called "correspondence" theory of truth." However this encounters serious problems. Philosophers, cognitive theorists and students of language have long thought that truth is "correspondence with fact" (*adequatio ad rem*). But on close inspection this emerges as a very questionable contention. Its tenability is problematic, its acceptability questionable. And what mainly renders it doubtful is the phenomenon of approximation.

Consider a simple illustration: what of yonder tree. Its age (so let us suppose) in actual fact 200 years. This given, it is duly false, and erroneous to say that it is roughly 240 years old. But what of saying that it is "roughly 200 years old?" This is true. But is it factual? In actuality trees do and must have a certain age. In reality's scheme of things, there is no such thing as a

---

identify a significant whole with 'an organized individual experience, self-filling and self-fulfilled'" (Joachim 1906, p. 78).

tree age or "roughly such-and-such." Truths can be roughly and approximate in ways to which reality cannot be accordant.

The actual age of a tree does not correspond to or identify with something rough, approximate, and indefinite. In sum truth can go where reality does not. Truth can be a matter of formulation in ways that reality does not match. Truths can be rough, approximate, and indefinite, but this is not an option for reality.

This point is brought out perhaps most vividly by disjunction. If—not knowing the difference—I say that the aversion I saw was either a camel of a dromedary I am speaking the truth. But there is no such thing as an either-or animal, and no disjunctive reality for my truth to characterize and correspond with.

No doubt truth is guided in reality and somehow consonant with it. But this guiding and this consonance is not a matter of adequation—of agreement or correspondence. For the reality of it is that the truth need not somehow describe reality.

That the average (typical) English seaman of Nelson's day was 5'1" 17 year old of 110lb. weight may well be true even if in fact no single such sailor answers to exactly this description. Such a claim can be adequately grounded in reality even when that reality is not a state of affairs that is somehow describe by it.

The grounding relationship that links truth to reality is something more complex and convoluted than anything that then terms description or correspondence or description manages to convey.

An Advantage of Coherentism—Generosity

Coherentism is clearly the more inclusive, far-reaching, and comprehensive standard. Unlike its rival fundamentalism it reaches beyond mere inferences from observational givens. It enables us to stake claims and answer questions over vast regions where its more restrictive alternative is condemned to silence. And its adherents see contextualism as too lightly bound to the apron-strings of observation. Consider a very simple example that if large sample theories and cosmological models. It is harmonization alone that enables us to project trends whose substantiation requires leaps across a chasm of imperfect information. The reach of theoretical science is simply longer.

An Advantage of Coherentism—Self-Support

Any effective inquiry method must conform its own strictures. A logic whose proceedings are not appropriately logical by its own telling, an arithmetic whose calculations are not justified by its own means, an epistemology whose proceedings fail to validate its own principles, must all be accounted as incomplete, imperfect, and inadequate. Any adequate cognitive method must be self-sustaining and appropriate on its own telling. In sum, any system rational internal procedure must be self-sustaining—it must on its own terms authorize and support the processes by which it itself proceeds.

Accordingly, an inquiry process for resolving issues of how things stand in the world must be able to address the reflexive question:

> Why and how is to that I (the method at issue) as able to resolve (be it correctly or at least roughly or approximately) questions about the nature of things? What is it about my modus operandi that makes me reliable in my own sight?

And this means that there must a process by which our own preferred inquiry process—the scientific method—is able to establish its own credentials, its claims to adequacy.

What we have here is demand for a cyctrically reflexive course of substantiation when fact-substantiating method substitutes our claims to fact substantiation and the facts at our disposal substitutes the efficacy of method. In this way, method and fact must be reciprocally reinforcing. This, however, is not a *vicious circle*. Rather, it is a *virtuous coordination*. If our thesis-substation process were not, on its own telling, the best-available option we have, then it would for that very reason be questionable and defective.

We both desire and require an inquiry process to be self-endorsing: to maintain on its own telling the claim that "I am the way to go at it."

Such self-endorsement is an indispensably necessary condition of adequacy. But is it also sufficient? Is textual support—endorsing the right claims and theses—enough to establish the adequacy of an inquiry process? Does it not, in the final analysis, require something more, something text-transcending that cannot be settled by merely accepting a story.

The answer is affirmative. Textual harmonization and implementation is needed: but it is *not* enough. For adequacy something further is required, something trans-textual, when at something that simply *describes* reality but *constitutes* it.

In prioritizing the systemic harmonizing of our cognitive commitment to coherentism embraces reflexivity by taking a self-supporting and self-sustaining position. It is clearly harmonious to take harmonization as the essential to the standard of cognitive acceptability.

An Advantage of Coherentism—Rational Unification

The information at our disposal that constitutes our putative knowledge renders two services for us. First, it provides answers to our questions about ourselves and our place in the world's scheme of things and thereby meets our intellectual needs for information. And second, it guides our actions in ways that meet our practical need for the realization of authentic (cognition-transcending) satisfactions. Foundationalism with its tethering to observation does only limited service in this regard. Coherentism with its broader perspective is in a position to meet our cognitive and practical needs on a more ample basis. Its adequacy to reality is not merely *observational* but *operational*, something that bears not simply on passive observations but on active procedure and interaction and construal.

An Advantage of Coherentism—Verisimilitude

Consider a gridwork of gridwork of the form

Figure 6.1: Gridwork

Where every successively central square is filled in exactly the same manner. In this way there would always be just as many light and dark squares overall. And yet *in appearance* it seems as though the dark predominates.

This sort of thing is typical of optical illusion such as the famous double arrow picture with its opposite-facing arrow-heads attached to line segments of the same length as per:

Figure 6.2: Arrows

It has long been evident that we need theory to correct experience—even the experience of our own senses. The entire context of relevant process is taken into account and made to harmonize overall, we are likely to have a firm grip on what actually obtains. In the interest of having cogent and instructive information about Reality we do well to spread our net for the accession and consolidation of information beyond strictly experiential, observationally substantiable limits.

The Pragmatic Upshot

Coherentism fuses two critical epistemic issues into a single unit—namely the consolidation or systematization of information and its verification or validation. Its procedure is to ground the acceptability of data is through their coherent coordination into a unified and systemically harmonious framework. Reciprocal fit and harmonization does double duty as the testing standard of acceptability, so that its test of acceptability is what might be called the Cinderella Standard: the proper shoe is the one that fits the best.

In spheres of inquiry when we cannot secure information without accepting contentions and where acceptance is always structured by the prospect of error, we have to take chances. And in this regard coherent is our best bet, not because it cannot lead to error, but because experience indicates that no alternative method affords better promise of success.

A Rationalistic Contextualism Is Not an Irrationalistic Indifferentism

The fact that the principles of probative reasoning in philosophy are themselves pre- or extra-evidential does not mean that they thereby lack a rational standing and stand benefit of a basis of rational justification. For they do indeed have such a basis in our experience. Only the general course of my experience can dictate whether and to what extent I actually have a plausible datum to work with. There is no doubt, however, that this makes for contextual variability. Different people with different bodies of experience will (quite "rationally") see the same evidence differently, and implement different standards of cognitive precedence and priority. Experience provides for variation with regard to what is admitted as datum, how this is to be weighed, and how this is to be used.

However, while human experience at large is diversified and variegated, any one particular individual's experience is unified through the very fact of its individualization. As Kant insisted, even as our shadow everywhere accompanies our body so the reflexive "I think" potentiality accompanies all our conscious thought. And while the complexity of experience makes for an aporetic inconsistency of beliefs the reflexive unity of experience also makes for determinateness on the side of epistemic evaluation, where normalcy and standardness emerge as statistical stabilities. Such factors may differ as between different people, but for the individual they come "one to a customer." There is no room left here for arbitrary "in the eyes of the beholder" differentiation.

There prevails a widespread tendency to see reason as limited to matters of inference and compatibility. David Hume wrote:

> It is not contrary to reason to prefer the destruction of the whole world to the scratching of my finger. It is not contrary to reason for me to choose my total ruin... It is as little contrary to reason to prefer even my own acknowledged lesser good to my greater, and have a more ardent affection for the former than the latter.[33]

But this is simply absurd in any sensible understanding of what reason is all about. Reason, after all, encompasses the area of *reasonableness* and the domain of what Kant called *judgment*. Any purely inferential view of reason goes severely amiss by overlooking issues of plausibility and tenability and consonance.

The issue of rational evaluation is not "a matter of taste" and its rulings do not "lie in the eyes of the beholder. In particular, although our cognitive values of normality coordinated precedence and priority "emerge from experience," they are not subjective (dependent on individual inclination) but variable (determined by the potentially different experimental situation of different individuals).

It is mistaken to characterize an experiential epistemology as a form of *relativism*. The contextualistic circumstance that different eras, different matters, and to some extent even different individuals are differently situated within the overall manifold of human experience it not a matter of relativistic ideolo-

---

33 *Treatise*, Bk. II, Pt. iii, sec 3.

gy—it is simply a fact of life. Coming to realistic terms with it is not a defect of objective rationality but an inevitable part of it.

Outright relativism is a mode of indifferentism: it sees the choice of a modus operandi as ultimately rational, as a subjective matter of mere taste or inclination—"You go your way and I go mine." The presently envisioned experiential coherentism is not a version of *relativism* but rather one of *contextualism*. It is deeply rationalistic, based on the idea that any rationally appropriate judgment must be based on the overall experience in hand. Rational persons align their judgments as closely as they can to the experience at their disposal. But different people exist in different experiential contexts, so that it would not only be unrealistic but actually absurd to expect the same diagnosis from a physician of Galen's day as from one of our own. Different contexts not only permit but actually demand different judgments from individuals. Contextualism is not a slide into relativism; it is a matter of good rationalistic common sense. After all, the underlying plausibility appraisal, at issue are—or should be—the product not of a matter of personal preference (tastes or inclinations) but objective appraisals of how individual items fit into the wider holistic structure of a person's overall experience. Coherentism in metaphilosophy is thus nowise a descent into subjectivistic relativism.

7 On Philosophical Systematization

The Hegelian Vision

In contemplating the great historical panorama of the history of philosophical thought in his *Vorlesungen über die Geschichte der Philosophie*, G. W. F. Hegel (1770–1831), conceived the interesting and ambitious prospect of encompassing the whole of it within the domain of a single, all-embracing system. In this way all of those diverse and conflicting theses and systems could be conjoined together as so many components of one unified whole of coordinated thought. All those seemingly rival systems with their conflicting ideas could be reconceptualized not as enemies-in-conflict but rather as cooperative constituents in the development of one vast and many-sided collaborative venture. The idea of one single grand all-encompassing system thus a siren song that lured Hegel and his followers down the primrose path of a dialectical syncretism. But this grand program had encounters substantial obstacles.

For Hegel, the history of philosophy is not a story of a random walk of mistaken efforts leaving a plethora of ruins along the way, but a progressive journey along a rationally evolving program of inquiry gradually unfolding a coherent system of thought in increasingly vivid details. As Hegel saw it, the course of philosophical history is an organic process of organic development through incompatible stages (à la chrysalis and butterfly) producing an increasingly complex end product in which those earlier are preserved and combined—albeit in a transfixed manner. The task of the serious student of philosophical history is thus not a matter of antiquarian compilation but one of philosophical creativity—to present the teachings of philosophers past in their conceptual and developmental role within a viable and coherent system of philosophical thought. As Hegel himself rather ambitiously put it: "The succession of philosophical systems in the course of history is one and the same as the succession of logical inference in the determination of the Idea."[34] The aim is not just to present the past as was, but to reveal how it has come to contribute to the manifold of philosophical truth as best we ourselves can manage to discern it. According-

---

Note: This essay was originally published in the *Southern Journal of Philosophy*, vol. 43(2005), pp. 425–442.

34 *Op. Cit.* Introduction.

ly, it is a prime task of philosophical inquiry is to accommodate, absorb, and coordinate the manifold of part philosophizing in systemic interaction.[35]

A Fatal Obstacle: The Plunge into Inconsistency

Easier said than done! For a fatal obstacle confronts any crude syncretism that seeks to conjoin or compile the diversity of historical projects in the field. For from the very outset, philosophy has been conceived of as a quest for truth. And the contentions and theories offered by various philosophers have been regarded by one and all who have contemplated them as so many truth-claims, though often as not as flawed ones.

But if this gearing of philosophy to truth is how one is going to proceed in construing philosophical claims, then the idea of a grand systemic compilation is destined to come to grief. For if you maintain that p and I maintain that not-p, then in conjoining our claims one has a logical contradiction on one's hands. And so any system that takes the route of a philosophical syncretism—of cognitively combining or pooling these claims—is not going to lead us to a larger or deeper view of things, but will end in a mess.

It has always been the goal of Hegel-inspired thinkers to treat philosophy-as-a-whole as a single system—a grand unified amalgamation of theses and ideas embracing the entity of thought on fundamental issues. Unfortunately for such an approach something very big and powerful has stood in the way, namely: logic. For the subject is, of course, replete with mutually incompatible contentions. And this very fact stands in the way of seeing the field as a manifold of truths. If the Aristotelian perspective is correct, and a system of rational cognition must—actually or potentially—constitute a truth-manifold, then philosophy-as-a-whole is ruled out of contention as a rational cognitive system from the very start. For to all appearances any fusionary project of combining or conjoining philosophical claims is going to issue in incoherence and incomprehension.

And has doubtless been for this reason that otherwise appealing idea of an historically all-absorbing system failed to meet with widespread sympathy among philosophers and students of the history of philosophy.

---

[35] A good compact account of Hegel's position regarding the history of philosophy is Braun 1973, pp. 328–40.

Plausibility to the Rescue

The roadblock to any mere compilation of philosophical doctrines is the result of the very problematic idea of treating those philosophical claims as so many rival truths. For it then follows at once that, since the truth-at-large is (and must be) collectively consistent while incompatible claims cannot all be true, the discordancy and inconsistency of philosophical contentions constitutes a decisive obstacle to a realization of the syncretist program.

There is, however, the prospect of a very different and decidedly more promising approach—albeit one that calls for a rational reconceptualization of the syncretist project at issue. For what if one were to embark on a seismic shift and undertake to reconceptualize the discipline not as a classically Aristotelian quest for truth but as a quest for plausibility instead? What if it were to take up the prospect of viewing philosophy construing philosophy-as-a-whole as a rational system of cognition alright, but a system geared at the explanation not of the manifold of accomplished truth but rather of the manifold of beckoning plausibility?

This approach calls for taking the very different line of approach based on the perfectly feasible prospect of viewing philosophical claims at large not as so many conflicting and incompatible truths, but rather as just so many coexisting plausibilities. This shift from truth to plausibility puts the whole matter in a very different light. For even as one can accept goods either "for keeps" or "on approval" so one can accept claims or contentions either by way of endorsement as true or by way of entertained as plausible.

But if you give up on truth, do you not thereby abandon the prospect of achieving something of cognitive significance? By no means! Plausibility after all is not nothing. It is, after all, closely linked to evidentiation, and while such evidentiation need not and often will not be decisive, the very fact that probability and plausibility are at issue prevents the matter from cognitive emptiness.

How Plausibility Works

In their single-minded focus upon truth and certainty, epistemologists have neglected the idea of plausibility. Plausibility is a matter of evidentiation. A proposition is plausible if some of the available evidence speaks significantly in its favor and the available evidence does not suffice to countervail against it. And so while truth/falsity is a matter of the status of a proposition in relation to the facts of the matter, its plausibility/implausibility is a matter of its status in relation to the state of its evidence—its relation to the *known* facts. If the extent to which the evidence speaks for something is not outweighed by available counterindications it straightaway passes the test of plausibility.

To be sure, plausibility is something very different from truth in its nature and in its logic since what is plausible need by no means be true. And there are various largely consonant ways of approaching the general idea of plausibility. First, from the angle of evidence: a proposition is plausible if there is some evidence for it and no *preponderating* evidence against it. Second, consider the situation from the vantage point of a specified spectrum of exhaustive but mutually exclusive possible cases. Then a proposition will be *true* in just exactly the one that is actual, it will be *possible* if it holds in any one of them, and it will be *plausible* if it holds in several (i.e., more than one) of them than its negation does. The third line of approach proceeds by way of probabilities. A proposition is *certain* if its probability equals on (= 1), it will be *possible* if its probability is greater than zero (> 0), and it will be *plausible* if its probability is nonminimal (say > 1/10).

Plausibility is thus something decidedly different from truth which, by its nature, must exhibit certain characteristic features of truth:

Conjunctivity: The conjunction of true propositions is true

If Tp and Tq, then $T(p \& q)$

Consistency: A proposition and its negation are never both true

Never: $T(p)$ and $T(\sim p)$

Transparency: One can see through the truth of a proposition to the proposition itself

If Tp, then p

However, none of these principles will invariably hold when we move from truth to plausibility, replacing the truth operator T by the plausibility operator P.

What this means is, in effect, that the domain of the plausible is ruled not so much by logical formalities as by substantive considerations, so that in dealing with plausibility one must be concerned for matters of substance as well as matters of form.

Plausibility Syncretism and Aporetics

The difference between truth and plausibility is crucial for preset purposes. For it opens up the prospect of a quasi-Hegelian system of philosophy, that is now conceived of not as a systematization of a domain of truth, but rather as a systematization of a domain of plausibility. There is, to be sure, little real prospect of classing the claims of rival philosophies as across-the-board *true*, seeing that once one sees one's own to be so, mere logic prevents doing owing to incompatibilities. But there is no logical obstacle to taking the rather generous (if perhaps overcautious) step of treating all the competing alternatives as so many (merely) plausible prospects. Plausibility just does not involve the same restrictive regimentation that logic imposes upon truth. The prospect of seeing rival philosophical positions as plausible and endeavoring to configure philosophy-as-a-whole in the fashion of a *plausibility* system (rather than a *truth* system) is not subject to any logical impediment.

A truth-syncretism makes no sense in philosophy, seeing that a manifold of logically inconsistent propositions is simply incoherent. But a plausibility-syncretism is something else again. In admitting inconsistencies at this level we do not blind the prospect of a rationally cogent systematization, but open the door to it—albeit by way of a plausibility rather than an unrealizable alethic (truth-geared) system.

Of course the result of such a compilation of discordant position is not a system of philosophy but rather a systematization of philosophizing at large. After all, as Kant already emphasized in his *Polegomena to any Future Metaphysics*, the study of philosophy is neither a method of nor a substitute for philosophizing itself.

When one takes the step of seeing philosophy-at-large in the light of a plausibility syncretism, one thereby transforms the study of philosophy into an exercise in aporetics.

An apory is a group of individually plausible propositions individually that are collectively inconsistent. In general if we gather together a cluster of contentions by philosophers on virtually any issue of the field, such a cluster will result. And the reason for the proliferation of apories in philosophy is simple. In this domain we are trying to answer large questions on the basis of small information. The evidence at our disposal is imperfect and incomplete in relation to the conclusions we have to draw from it. Almost unavoidably the projection of what we have into what we need can be made in different and incompatible directions. In moving from the given to the plausible the prospect of aporetic conflict is pretty well inevitable.

As an example, the theory of knowledge of the ancient Greeks revolved about the following quartet of currently mooted albeit collectively incompatible contentions:
1. We do have some knowledge about the world;
2. Whatever knowledge we have about the world must come via the senses (i.e., ultimately roots in what the senses deliver);
3. There is no genuine knowledge *(episteme)* without certainty;
4. The senses do not yield certainty.

Any positive inclination toward these theses—any tendency to see them each as plausible and (presumptively) acceptable—sets the stage for philosophical conflict. A (limited) variety of exits from the inconsistency is available:
- *Deny 1:* Maintain that we cannot have authentic knowledge about the world (the Pyrrhonian Sceptics).
- *Deny 2:* Maintain that genuine knowledge about the world can come from reason alone (Pythagoreans, Plato).
- *Deny 3:* Maintain that adequate knowledge need pot be based on the certain but can be based on the plausible-to *pithanon* (the Academic Sceptics).
- *Deny 4:* Maintain that the senses do yield certainty in some cases-those that result in the so-called "cataleptic" perceptions (the Stoics).

Faced with that aporetic cluster, we must make up our minds to decide between these alternatives.

Again, consider the following "Cartesian" apory regarding the nature of knowledge.
1. Knowledge must be absolutely certain;
2. Absolute certainty is never available with matters of fact;
3. Factual knowledge is available.

Here we have a straightforward example of an inconsistent triad. Descartes and the skeptics alike agree on (1). But in the interest of consistency, Descartes sacrificed (2) to (3) while his skeptical predecessors in classical antiquity sacrificed (3) to (2). Clearly one cannot maintain all three propositions as conjointly true. Any such family of plausible-seeming but inconsistent contentions sets the stage of an aporetic situation that cries out for resolution.

Apory Engenders a Diversity of Resolutions

Whenever we are confronted with an aporetic cluster, a plurality of resolutions is always available. The contradiction that arises from overcommitment can be resolved by abandoning any of several contentions, so that alternative ways of averting inconsistency can always be found.

Thus consider the theory of morality developed in Greek ethical thought affords a good example of such an aporetic situation. Greek moral thinking inclined to the view that the distinction between right and wrong:
1. Does matter;
2. Is based on custom *(nomos)*;
3. Can only matter if grounded in the objective nature of things *(phusei)* rather than in mere custom.

Here too an aporetic problem arises. The inconsistency of these contentions led to the following resolutions:
- *Deny 1:* Issues of right and wrong just don't matter-they are a mere question of power, of who gets to "lay down the law" (Thrasymachus);
- *Deny 2:* The difference between right and wrong is not a matter of custom but resides in the nature of things (the Stoics);
- *Deny 3:* The difference between right and wrong is only customary *(nomoi)* but does really matter all the same (Heraclitus).

We have here a paradigmatic example of an antinomy: a *theme* provided by an aporetic cluster of propositions, with *variations* set by the various ways of resolving this inconsistency. The problem of the philosopher is not one of inductive *amplification* but of systemic reduction of a restoration of consistency. And philosophers fail to reach a uniform result because this objective can always be accomplished in very different ways.

If we have firm confidence in our reasonings, then it follows by the inferential principle of *modus tollens* that whenever a belief is rejected, one must also call into question some of the various (collectively compelling) reasons on whose basis this belief had been adopted. For example, if one rejects free will, then one must also reject one of the following (presumptive) initial reasons for espousing freedom of the will: "People are usually responsible for their acts," "People are only morally responsible for those acts that are done freely." The rejection of an accepted thesis at once turns the family of reasons for its adoption into an aporetic cluster. Apory, once present, tends to spread like wildfire through any rational system.

This line of consideration accounts for what is, on first view, a puzzling aspect of the field, namely, the prominence in the philosophical literature of counter-argumentation and refutatory discussions. In mathematics no one troubles to argue that fourteen or thirty-two is *not* a satisfactory solution to a certain problem. This would be pointless because the number of incorrect answers is endless. But when there is only a limited number of viable alternative candidates in the running, negative and eliminative argumentation will obviously come to play a much more substantial part.

The Greek theory of virtue affords another example:
1. If virtue does not produce happiness/ pleasure, then it is pointless;
2. Virtue is not pointless-indeed it is extremely important;
3. Virtue does not always yield happiness.

Three ways of averting inconsistency are available here:
- *Deny 1:* Maintain that virtue is worthwhile entirely in itself, even if it does not produce happiness/pleasure (Stoics, Epictetus, Marcus Aurelius);
- *Deny 2:* Maintain that virtue is ultimately pointless and can be dismissed as folly of the weak (nihilistic sophists, e.g., Plato's Thrasymachus);
- *Deny 3:* Maintain that virtue is automatically bound to produce happiness (of itself always yields *real* pleasure)-so that the two are inseparably interconnected (Plato, the Epicureans).

The whole of the group (1)–(3) represents an aporetic cluster that reflects a cognitive overcommitment. And this situation is typical: the problem context of philosophical issues standardly arises from a clash among individually tempting but collectively incompatible overcommitments. Philosophical issues standardly center about an aporetic cluster of this sort-a family of plausible theses that is assertorically *overdeterminative* in claiming so much as to lead into inconsistency.

To put matters to rights, in such cases, something obviously has to go. Whatever favorable disposition there may be toward these plausible theses, they cannot be maintained in the aggregate. We are confronted by a (many-sided) cognitive dilemma and must find one way out or another. In particular, we can proceed:
- To reason from (2)-(3) to the denial of (1);
- To reason from (1), (3) to the denial of (2);
- To reason from (1)-(2) to the denial of (3).

An apory gives rise to a group of valid arguments leading to mutually contradictory conclusions, yet each having only plausible theses as premises. It is clear in such cases *that* something has gone amiss, though it may well be quite unclear just where the source of difficulty lies.

It lies in the logical nature of things that there will always be multiple exits from aporetic inconsistency. For whenever such an antinomy confronts us, then no matter which particular resolution we ourselves may favor, and no matter how firmly we are persuaded of its merits, the fact remains that there will also be other, alternative ways of resolving the inconsistency. For a contradiction that arises from over-commitment can always be averted by abandoning various subgroups among the conflicting contentions, so that distinct awareness to averting inconsistency can always be found. As far as abstract rationality goes, alternative resolutions always remain open—resolutions leading to mutually contrary and inconsistent results. An aporetic cluster is thus an invitation to conflict: its resolution will only one of a coordinated group of mutually discordant doctrines (positions, teachings, *doxa*). The cluster accordingly sets the stage for divergent "schools of thought" and provides the bone of contention for an ongoing controversy among them. In philosophy, any family of inconsistent theses spans a "doctrinal spectrum" that encompasses a variety of interrelated albeit incompatible positions.

Aporetic Antinomies Structure the Issues

It emerges against this background how it is that an aporetic perspective or philosophizing comes to be significantly instructive:
- We now see these propositions in their interrelational interconnectedness. We come to realize that they are related notwithstanding the prospect of a radical diversity of thematic subject matter;
- We are confronted in a very clear and urgent way with the need for choice insofar as it is truth that is our goal;
- We get a clear view of the battlefield—and are able to pinpoint with enhanced precision and detail exactly where the discordances between alternative parties are located.

Consider an example. The theory of morality developed in Greek ethical thought, which affords a good illustration of such an aporetic situation, was based on three plausible considerations:
- If virtue does not produce happiness, pleasure, then it is pointless;
- Virtue is not pointless—indeed it is extremely important;
- Virtue does not always yield happiness.

These, however, are collectively inconsistent. And three ways of averting inconsistency are available here:
- *Deny 1:* Maintain that virtue is worthwhile entirely in itself, even if it does not produce happiness/pleasure (Stoics, Epictetus, Marcus Aurelius);
- *Deny 2:* Maintain that virtue is ultimately pointless and can be dismissed as a folly of the weak (nihilistic sophists, e. g., Plato's Thrasymachus);
- *Deny 3:* Maintain that virtue is automatically bound to produce happiness (of itself always yields *real* pleasure)—so that the two are inseparably interconnected (Plato, the Epicureans).

We have here a paradigmatic example of an antinomy: a *theme* provided by an aporetic cluster of propositions, with *variations* arising from the various ways of resolving this inconsistency.

And this example illustrates any particular way out of an aporetic conflict is bound to be simply *one way among others*. The single most crucial fact about an aporetic cluster is that there will always be a variety of distinct ways of averting the inconsistency into which it plunges us. And in this light, the problem for the philosopher is not one of inductive amplification but of systemic reduction—of a restoration of consistency through choices of priority. In general, to be sure, philosophers fail to reach a uniform result because this prioritization can in theory always be accomplished in very different ways. The crux is that different philosophers implement different priority systems in effecting such determinations about what must be made to give way. Any and every resolution of a philosophical antinomy represents a distinct—and distinctly different—position, an intellectual abode that someone caught up in the underlying apory may choose to inhabit, though sometimes no one does so.

The state of affairs we have been considering stands in an interesting and ironic contrast with that of Plato, that greatest of philosophers. He taught that *sensation* yields contradictory results and leads to belief (*pistis*), whose "object can be said both to be and not to be" (*Republic*, Bk. V, 478). (Think of the sceptics' favorite example of the two hands, one held in hot water and the other in cold water, and then both plunged into lukewarm water.) And such incoherence means that sensory beliefs must be corrected by *dianoia*—by reason. As Plato thus saw it, the philosopher's theorizing is the saving resource capable of effecting a reconciliation between the conflicting data of sensory observation.

Their grounding in aporetic conflicts provides philosophical controversies with a natural structure that endows its problem areas with an organic unity. The various alternative ways of resolving such a cognitive dilemma present a restricted manifold of interrelated positions—a comparatively modest inventory of possibilities mapping out a family of (comparatively few) alternatives that span the entire spectrum of possibilities for averting inconsistency.[36] And the history of philosophy is generally sufficiently fertile and diversified that all the alternatives—all possible permutations and combinations for problem resolution—are in fact tried out somewhere along the line.

Philosophical doctrines are accordingly not discrete and separate units that stand in splendid isolation. They are articulated and developed in reciprocal interaction. But their natural mode of interaction is *not* by way of mutual supportiveness. (How could it be, given the mutual exclusiveness of conflicting doctrines?) Rather, competition and controversy prevail. The search of the ancient Stoics and Epicureans (notably Hippias) for a universally "natural" belief system based on what is common to different groups (espousing different doctrines, customs, moralities, religions) is of no avail because no single element remains unaffected as one moves across the range of variation. Given that rival "schools" resolve an aporetic cluster in different and discordant ways, the area of agreement between them, though always there, is bound to be too narrow to prevent priorities are by nature incompatible and irreconcilable.

Other illustrations are readily available. A metaphysical determinism that negates free will runs afoul of a traditionalistic ethical theory that presupposes it. A philosophical anthropology that takes human life to originate at conception clashes with a social philosophy that sees abortion as morally unproblematic. A theory of rights that locates all responsibility in the contractual reciprocity of freely consenting parties creates problems for a morality of concern for animals. And the list goes on and on.

Dialectics a Mechanism of System Growth and Development and the Role of Distinctions

One important insight that a resort to plausibility aporetics puts at our disposal relates to its revelation of developmental dialectics.

To be sure, Aristotle was right in saying that philosophy begins in wonder and that securing concerns to our questions is the aim of the enterprise. But of course we do not just want answers but coherent answers, seeing that these alone have a chance of being collectively true. The quest for consistency is an

---

36 This general position that philosophical problems involve antinomic situations from which there are only finitely many exits (which, in general, the historical course of philosophical development actually indicates) is foreshadowed in the deliberations of Wilhelm Dilthey 1961 (see p. 138).

indispensable part of the quest for truth. The quest for consistency is one of the driving dynamic forces of philosophy.

But the cruel fact is that theorizing itself yields contradictory results. In moving from empirical observation to philosophical theorizing, we do not leave contradiction behind—it continues to dog our footsteps. And just as reason must correct sensation, so more refined and elaborate reason is always needed as a corrective for less refined and elaborate reason. The source of contradiction is not just in the domain of sensation but in that of reasoned reflection as well. We are not just *led into* philosophy by the urge to consistency, we are ultimately *kept* at it by this same urge.

Accordingly, aporetics is not just a mapping the cartography of the battlefield of philosophical disputation but also a tool for understanding and explaining the dialectic of historical development. For in breaking out of the cycle of inconsistency created by an aporetic cluster one has no choice but to abandon one or the other of the propositions involved. But in jettisoning this item it is often—perhaps even generally—possible to embody a distinction that makes it possible to retain something of what is being abandoned. Consider the following example:

1. Every occurrence in nature is caused;
2. Causes necessitate their consequences;
3. Necessitation precludes contingency;
4. Some occurrences in nature are contingent.

Someone who decides to break the cycle of inconsistency by dropping thesis (3) might nevertheless causal necessitation and causal production and in consequence maintain that what causes do not necessitate these effects they may nevertheless produce them (albeit in ways that are not at issue with the contingency of product).

To restore consistency among incompatible beliefs calls for abandoning some of them as they stand. In general, however, philosophers do not provide for consistency-restoration wholly by way of rejection. Rather, they have recourse to *modification*, replacing the abandoned belief with a duly qualified revision thereof. Since (by hypothesis) each thesis belonging to an aporetic cluster is individually attractive, simple rejection lets the case for the rejected thesis go unacknowledged. Only by modifying the thesis through a resort to distinctions can one manage to give proper recognition to the full range of considerations that initially led into aporetic difficulty.

Distinctions enable the philosopher to remove inconsistencies not just by the brute negativism of thesis *rejection* but by the more subtle and constructive device of thesis *qualification*. The crux of a distinction is not mere negation or de-

nial, but the amendment of an untenable thesis into something positive that does the job better. By way of example, consider the following aporetic cluster:
1. All events are caused;
2. If an action issues from free choice, then it is causally unconstrained;
3. Free will exists—people can and do make and act upon free choices.

Clearly one way to exit from inconsistency is to abandon thesis (2). We might well, however, do this not by way of outright abandonment but rather by speaking of the "causally unconstrained" only in Spinoza's manner of *externally* originating casualty. For consider the result of deploying a distinction that divides the second premise into two parts:

2.1 Actions based on free choice are unconstrained by *external* causes;
2.2 Actions based on free choice are unconstrained by *internal* causes.

Once (2) is so divided, the initial inconsistent triad (1)–(3) give way to the quartet (1), (2.1), (2.2), (3). But we can resolve *this* aporetic cluster by rejecting (2.2) while yet retaining (2.1)—thus in effect *replacing* (2) by a weakened version. Such recourse to a distinction—here that between internal and external causes—makes it possible to avert the aporetic inconsistency and does so in a way that minimally disrupts the plausibility situation.

To examine the workings of this sort of process somewhat further, consider an aporetic cluster that set the stage for various theories of early Greek philosophy:
1. Reality is one (homogeneous);
2. Matter is real;
3. Form is real;
4. Matter and form are distinct sorts of things (heterogeneous).

In looking for a resolution here, one might consider rejecting (2). This could be done, however, not by simply *abandoning* it, but rather by *replacing* it—on the idealistic precedent of Zeno and Plato—with something along the following lines:

2'. Matter is not real as an independent mode of existence; rather it is merely quasi-real, a mere *phenomenon*, an appearance somehow grounded in immaterial reality.

The new quartet (1), (2'), (3), (4) is entirely contenable.

Now in adopting this resolution, one again resorts to a *distinction*, namely that between
- Strict reality as self-sufficiently independent existence;

and
- Derivative or attenuated reality as a (merely phenomenal) product of the operation of the unqualifiedly real.

Use of such a distinction between unqualified and phenomenal reality makes it possible to resolve an aporetic cluster—yet not by simply *abandoning* one of those paradox-engendering theses but rather by *qualifying* it. (Note, however, that once we follow Zeno and Plato in replacing (2) by (2′)—and accordingly reinterpret matter as representing a "mere phenomenon"—the substance of thesis (4) is profoundly altered; the old contention can still be maintained, but it now gains a new significance in the light of new distinctions.)

Again one might—alternatively—abandon thesis (3). However, one would then presumably not simply adopt "form is not real" but rather would go over to the qualified contention that "form is not *independently* real; it is no more than a transitory (changeable) state of matter." And this can be looked at the other way around, as saying "form *is* (in a way) real, although only insofar as it is taken to be no more than a transitory state of matter." This, in effect, would be the position of the atomists, who incline to see as implausible any recourse to mechanisms outside the realm of the material.

Aporetic inconsistency can always be resolved in this way; we can always "save the phenomena"—that is, retain the crucial core of our various beliefs in the face of apparent consideration—by introducing suitable distinctions and qualifications. Once apory breaks out, we can thus salvage our philosophical commitments by *complicating* them, through revisions in the light of appropriate distinctions, rather than abandoning them altogether.

The exfoliative development of philosophical systems is driven by the quest for consistency. Once an apory is resolved through the decision to drop one or another member of the inconsistent family at issue, it is only sensible and prudent to try to salvage some part of what is sacrificed by introducing a distinction. Yet all too often inconsistency will break out once more within the revised family of propositions that issues from the needed readjustments. And then the entire process is carried back to its starting point so that the over-all course of development thus exhibits an overall cyclical structure.

The unfolding of distinctions has important ramifications in philosophical inquiry. As new concepts crop up in the wake of distinctions, new questions arise regarding their bearing on the issues. In the course of securing answers to our old questions we open up further questions, questions that could not even be asked before.

The historical course thus tracks an evolving process of apory resolution by means of distinctions. And this process of dialectical development imposes certain characteristic structural features upon the course of philosophical history:
- Concept proliferation—ever more elaborate concept manifolds evolve;
- Concept sophistication—ever more subtle and fine-drawn distinctions;
- Doctrinal complexification—ever more extensively formulated theses and doctrines;
- System elaboration—ever more elaborately articulated systems.

However, this generic characterization of the matter does not do adequate justice to how things actually work. To improve matters it is advisable to look at some actual "real-life" examples from the history of philosophy.

The history of philosophy is shot through with distinctions introduced to avert aporetic difficulties. Already in the dialogues of Plato, the first systematic writings in philosophy, we encounter distinctions at every turn. In Book I of the *Republic*, for example, Socrates' interlocutor quickly falls into the following apory:
1. Rational people always pursue their own interests:
2. Nothing that is in a person's interest can be disadvantageous to him;
3. Even rational people sometimes do things that prove disadvantageous.

Here, inconsistency is averted by distinguishing between two senses of the "interests" of a person—namely what is *actually* advantageous to him and what he merely *thinks* to be so, that is, between *real* and *seeming* interests. Again, in the discussion of "nonbeing" in the *Sophist*, the Eleatic stranger entraps Theaetetus in an inconsistency from which he endeavors to extricate himself by distinguishing between "nonbeing" in the sense of not existing *at all* and in the sense of not existing *in a certain mode*. For the most part, the Platonic dialogues present a dramatic unfolding of one distinction after another.

And this situation is typical in philosophy. The natural dialectic of problem solving here drives us even more deeply into drawing distinctions, so as to bring new, more sophisticated concepts upon the scene.

To be sure, distinctions are not needed if *all* that concerns us is averting inconsistency; simple thesis abandonment, mere refusal to assert, will suffice for that end. One can guard against inconsistency by avoiding commitment. But such sceptical refrainings create a vacuum. Distinctions are indispensable instruments in the (potentially never-ending) work of rescuing the philosopher's assertoric commitments from inconsistency while yet salvaging what one can. They become necessary if we are to maintain informative positions and provide answers to our questions. Whenever a particular aporetic thesis is rejected, the optimal course is not to abandon it altogether, but rather to minimize the loss by introducing a distinction by whose aid it may be retained *in part*. After all, we do have some commitment to the data that we reject, and are committed to saving as much as we can. (This, of course, is implicit in our treating those data as such in the first place.)

A distinction accordingly reflects a *concession*, an acknowledgment of some element of acceptability in the thesis that is being rejected. However, distinctions always bring a new concept upon the stage of consideration and thus put a new topic on the agenda. And they thereby present invitations to carry the discussion further, opening up new issues that were heretofore inaccessible. Distinctions are the doors through which philosophy moves on to new questions and problems. They bring new concepts and new theses to the fore.

Distinctions enable us to implement the irenic idea that a satisfactory resolution of aporetic clusters will generally involve a compromise that somehow makes room for all parties to the contradiction. The introduction of distinctions thus represents a Hegelian ascent—rising above the level of antagonistic doctrines to that of a "higher" conception, in which the opposites are reconciled. In introducing the qualifying distinction, we abandon that initial conflict-facilitating thesis and move toward its counterthesis—but only by way of a duly hedged synthesis. In this regard, distinction is a "dialectic" process. This role of distinctions is also connected with the thesis often designated as "Ramsey's Maxim." For with regard to disputes about fundamental questions that do not seem capable of a decisive settlement, Frank Plumpton Ramsey wrote: "In such cases it is a heuristic maxim that the truth lies not in one of the two disputed views but in some third possibility which has not yet been thought of, which we can only discover by rejecting something assumed as obvious by both the disputants."[37] On this view, then, distinctions provide for a higher synthesis of opposing views; they prevent thesis abandonment from being an *entirely* negative process, affording us a way of salvaging something, of giving credit where credit is due" even to those theses we ultimately reject. They make it possible to remove inconsistency not just by the brute force of thesis rejection, but by the more subtle and constructive device of thesis qualification.

Philosophical distinctions are thus creative innovations. There is nothing routine or automatic about them—their discernment is an act of inventive ingenuity. They do not elaborate preexisting ideas but introduce new ones. They not only provide a basis for understanding better something heretofore grasped imperfectly but shift the discussion to a new level of sophistication and complexity.

---

[37] Ramsey 1931, pp. 115–16.

Thus, to some extent they "change the subject." (In this regard, they are like the conceptual innovations of science which revise rather than explain prior ideas.)

Philosophy's recourse to ongoing conceptual refinement and innovation means that a philosophical position, doctrine, or system is never closed, finished, and complete. It is something organic, every growing and ever changing—a mere tendency that is in need of ongoing development. Its philosophical "position" is never actually that—it is inherently unstable, in need of further articulation and development. Philosophical systematization is a process whose elements develop in stages of interactive feedback—its exfoliation is a matter of dialectic, if you will.

An Historical Illustration

The unfolding of distinctions accordingly plays a key role in philosophical inquiry because new concepts crop up in their wake so as to open up new territory for reflection. In the course of philosophy's dialectical development, new concepts and new theses come constantly to the fore and operate so as to open up new issues. And so in securing answers to our old questions we come to confront new questions that could not even be asked before.

The inherent dynamic of this dialectic deserves a closer look. Let us consider an historical example. The speculations of the early Ionian philosophers revolved about four theses:
1. There is one single material substrate (*archê*) of all things;
2. The material substrate must be capable of transforming into anything and everything (and thus specifically into each of the various elements);
3. The only extant materials are the four material elements: earth (solid), water (liquid), air (gaseous), and fire (volatile);
4. The four elements are independent—none gives rise to the rest.

Different thinkers proposed different ways out of this apory:
- Thales rejected (4) and opted for water as the *archê*;
- Anaximines rejected (4) and opted for air as the *archê*;
- Heraclitus rejected (4) and opted for fire as the *archê*;
- The Atomists rejected (4) and opted for earth as the *archê*;
- Anaximander rejected (3) and postulated an indeterminate *apeiron*;
- Empedocles rejected (1), and thus also (2), holding that everything consists in *mixtures* of the four elements.

Thus virtually all of the available exits from inconsistency were actually used. The thinkers involved either resolved to a distinction between genuinely primacy and merely derivative "elements" or, in the case of Empedocles, stressed the distinction between mixtures and transformation. But all of them addressed the same basic problem—albeit in the light of different plausibility appraisals.

As the Presocratics worked their way through the relevant ideas, the following conceptions came to figure prominently on the agenda:

(I)
- (1) Whatever is ultimately real persists through change.
- (2) The four elements—earth (solid), water (liquid) air (gaseous), and fire (volatile)—do not persist through change as such.
- (3) The four elements encompass all there is by way of extant reality.

Three basic positions are now available:
- *Abandonment 1:* Nothing persists through change—*panta rhei*, all is in flux (Heraclitus);
- *Abandonment 2:* One single elements persists through change—it alone is the *archê* of all things; all else is simply some altered form of it. This uniquely unchanging element is: earth (atomists), water (Thales), air (Anaximines). Or again, *all* the elements persist through change, which is only a matter of a variation in mix and proportion (Empedocles);
- *Abandonment 3:* Matter itself is not all there is—there is also its inherent geometrical structure (Pythagoras) or its external arrangement in an environing void (atomists). Or again, there is also an immaterial motive force that endows matter with motion—to wit, "mind" (*nous*) (Anaxagoras).

Let us follow along in the track of atomism by abandoning (3) though the distinction between material and non-material existence. With this cycle of dialectical development completed, the following aporetic impasse arose in pursuing the line of thought at issue:

(II)
- (1) Change really occurs.
- (2) Matter (solid material substance) does not change, nor does vacuous emptiness.
- (3) Matter and the vacuum are all there is.

As always, different ways of escaping from contradiction are available:
- *Abandonment 1:* Change is an illusion (Parmenides, Zeno, Eleatics);
- *Abandonment 2:* Matter (indeed *everything*) changes (Heraclitus);

- *Abandonment 3:* Matter and the word are not all there is; there is also the void—and the changing configurations of matter within it (atomism).

Taking up the third course, let us continue to follow the atomistic route. Note that this does not *just* call for abandoning (3), but also calls for sophisticating (2) to

> 2′. Matter as such is *not* changeable—it only changes in point of its variable rearrangements.

The distinction between *positional* changes and *compositional* changes comes to the fore here. This line of development has recourse to a "saving distinction" by introducing the new topic of variable configurations (as contrasted with such necessary and invariable states as the shapes of the atoms themselves).

To be sure, matters do not end here. A new cycle of inconsistency looms ahead. For this new topic paves the way for the following apory:

(III) $\begin{cases} (1) \text{ All possibilities of variation are actually realized.} \\ (2) \text{ Various different world arrangements are possible.} \\ (3) \text{ Only one world is real.} \end{cases}$

Again different resolutions are obviously available here:
- *Rejection 1:* A theory of real chance (*tuchê*) or contingency that sees various possibilities as going unrealized (Empedocles);
- *Rejection 2:* A doctrine of universal necessitation (the "block universe" of Parmenides);
- *Rejection 3:* A theory of many worlds (Democritus and atomism in general).

As the atomistic resolution represented by the second course was developed, apory broken out again:

(IV) $\begin{cases} (1) \text{ Matter as such never changes—the only change it admits of are its rearrangements.} \\ (2) \text{ The nature or matter is indifferent to change. Ist rearrangements are contingent and potentially variable.} \\ (3) \text{ Its changes of condition are inherent in the (unchanging) nature of matter—they are necessary, not contingent.} \end{cases}$

Here the orthodox atomistic solution would lie in abandoning (3) and replacing it with

3′. Its changes of condition are not necessitated by the nature of matter. They are indeed quasi-necessitated by being law determined, but law is something independent of the nature of matter.

The distinction between internally necessitated changes and externally and accidentally imposed ones enters upon the scene. This resolution introduces a new theme, namely *law determination* (as introduced by the Stoics).

Yet when one seeks to apply this idea it seems plausible to add:

(V)4. Certain material changes (contingencies, concomitant with free human actions) are not law determined

Apory now breaks out once more; the need for an exit from inconsistency again arises. And such an exit was afforded by (4)-abandonment, as with the law abrogation envisaged in the notorious "swerve" of Epicurus, or by (3′)-abandonment, as with the more rigoristic atomism of Lucretius.

The developmental sequence from (I) through (V) represents an evolution of philosophical reflection through successive layers of aporetic inconsistency, duly separated from one another by successive distinctions. This process that led from the crude doctrines of Ionian theorists to the vastly more elaborate and sophisticated doctrines of later Greek atomism.

And this historical illustration indicates an important general principle. The continual introduction of the new ideas that arise in the wake of new distinctions means that the ground of philosophy is always shifting beneath our feet. And it is through distinctions that philosophy's prime mode of innovation—namely *conceptual* innovation—comes into play. And those novel distinctions for our concepts and contexts for our theses alter the very substance of the old theses. The dialectical exchange of objection and response constantly moves the discussion onto new—and increasingly sophisticated—ground. The resolution of antinomies through new distinctions is thus a matter of creative innovation whose outcome cannot be foreseen.

A Retrospect to Herbart

A dialectical process of Hegelian proportions is at work throughout aporetic dialectics. According to Hegel. it is the essential character of human reason to involve itself u, contradictions-conflicts of commitment that it first posits but then overcomes through an eventual reconciliation at a higher level. However, the philosopher who analyzed this aspect of the history of the subject most clearly was Johann Friedrich Herbart Hegel's younger contemporary (1776–1841). He proposed that the history of philosophy should be recast in issue-oriented form and should in fact be written in terms of the development of doctrines devised to resolve successively encountered antinomies. The history of philosophy, he held, should be written as a history of problems (and thus in a genre of which, even today, we have but a few fragmentary samples).

Herbart maintained that the fundamental concepts of thing/substance (as a unification of a plurality of different and distinct qualities is one single item) and of causality (as the production by one item or state of as yet another that is substantially different) are at bottom conceptually inconsistent in forcing discordant factors into a logically unwarrantable unity. And the same goes for the idea of a self as the unitary basis of diversified doings. Logic can only underwrite connections of necessity: contingent connections are beyond its rank—literally incoherent from a logic-conceptual standpoint. The endeavor of philosophical theorists to impose a neat logico-conceptual order on a fundamentally surd and contingent reality is bound to issue in aporetic disunity.

As Herbert saw it, the experientially grounded concepts in whose terms we represent and process our cognitive experiences in science and ordinary life always involve internal conflicts. An experiential concept A unites two disparate elements M and N that do not stand in a logico-conceptual union but are united by a strictly factual bond. There is a tension or contradiction here. We can neither (on theoretical grounds) maintain that there is, a fusion of M and N in A, nor yet (on factual grounds) can we deny this connection outright. Logic rejects the conceptual fusing of M and N. experience rejects their separation. All we can do is suppose that there is some new element, some distinction that splits M into M_1 and M_2 one of which is rigidly joined to N. the other strictly distinct from it At best, then, we can see A as an unstable compound, oscillating between A_1 (where M_1 is problematically conjoined with N) and A_2 (where M_2 is unproblematically disjoined from N). Accordingly, every experiential concept is the ground from which some suitable "supplementary concept must emerge to yield a distinction capable of restoring consistency.

Herbart saw the prime task of philosophy as the reworking of our experiential concepts so as to restore consistency—to effect an integration that relegates these inner contradictions to the realm of mere appearance. Philosophy strives to overcome the internal inconsistency of our presystemic concepts. Throughout our philosophizing, those experiential concepts will inevitably come to be transcended by successors who seek to resolve the tensions of their presystemic predecessors. This process, Herbart's "method of relations" (*Methode der Beziehungen*), is the counterpart in his system of the Hegelian dialectic. As Wilhelm Dilthey put it:

Herbart was the first who regressed analytically from the course of philosophical development to the particular problems that were the prime mover in the minds of individual thinkers. For him, philosophy was "the systematic study (*Wissenschaft*) of philosophical questions and problems." And so he responded to the question of the nature of philosophizing with the reply that it is "the endeavor to solve problems." In the first redaction of his *Introduction to Philosophy*, he places the motive force to philosophizing in the puzzles and contradictions regarding the nature of things. Our trying to put the pieces together, to see the world whole, occasions our initial discovery of philosophical problems.[38]

Herbart thus deserves to rank along with Hegel as a founder of the theory of aporetic dialectic in philosophy.

Philosophy in a Different Light: Recovering the Hegelian Vision of Philosophy at Large

The turn to plausibility open up different way of viewing philosophy-at-large and of organizing the history of philosophy on rational (or at least more perspicuous) principles.

To be sure, in looking to our own philosophy we are, of course, minded to see its various contentions as truths—and thereby see the rival alternatives as falsehoods and errors. But there is also a somewhat more generous prospect. For one has the option of regarding *the entire manifold of the contentions of philosophers*—ourselves included—*as so many (merely) plausible propositions*.

This of course involves a radical departure from the all too common way of looking at philosophy-at-large namely as a deeply flawed venture in the quest for truth—resulting in a mixed bag that conjoins some a small aggregate of truths (one's own views) along with a massive plurality of error (everyone else's). If our concept of cognitive systematization is confined to the classical, Aristotelian view, then there just is no possibility of systematizing philosophy-as-a-whole. But according to philosophical contentions the status of plausibilities we open up the prospect of systematizing philosophy at large and one single and unified—albeit vast—non-Aristotelian system of rational cognition.

An approach which see philosophical contentions as ("merely") plausible thus open up the prospect of regarding philosophy-at-large as a meaningful venture in rational cognition—one of constructing a non-Aristotelian system of *plausible* responses to the big questions comprised in the problem-agenda of the field. And in thus articulating a philosophical apory and elaborating the possi-

---

38 Dilthey 1961, p. 134.

bilities for its resolution, and then exfoliating the plausible save-what-you-can distinctions we are, in effect, spelling out some component sector of the large system constitution a non-Aristotelian systematization of philosophy-at-large.

In turning from truth to plausibility one realizes both gains and losses. The gains relate to amplitude of vision and breadth of perspective: a great many more things are plausible than are determinately true. The loss relates to reliability: a good deal of shaky stuff gets added in and there will be more dubious dross amid the reliable gold. Accordingly, in interpreting philosophy-at-large in the light of plausibility considerations one takes a distinctive and in some ways non-doctrinal line.

For while philosophy is often characterized as a quest for truth, this strategy realizes the prospect of an entirely different approach to information—one geared not to irrefragable truth but the fallible plausibility.

Such an approach represents a vision that has been on the stage in German philosophy since Christian Wolff[39] and prominent since Hegel. And it was in this frame of mind that the Bertrand Russell of the pre-World War I era wrote: "Philosophy is to be studied, not for the sake of any definite answers to its questions, since no definite answers can, as a rule, be known to be true, but rather for the sake of the questions themselves; because these questions enlarge our conception of what is possible, extend our intellectual imagination, and diminish the dogmatic assurance which closes the mind against speculation."[40]

It must be emphasized, however, that in taking such a more inclusive and many sided view of the subject, we are in process of addressing philosophy at large, and not deriving our own philosophy. We survey, examine, and weigh the size of possible answers to the questions with their deciding which one to accept as correct. We are, in sum, looking at the matter from the standpoint of the community, not from that of ourselves in *propria persona*.

For, what we get in such a quasi-Hegelian perspective is not a system of philosophy—not a coherent and cohesive exposition of a philosophical position that offers specific answers to definite questions. Rather what we get is a systematization of philosophizing-at-large, a comprehensive coordination of philosophizing in general. After all, a plausibility system, unlike a system of purported truth, does not provide *an answer* to any of the questions—or *a solution* to any of the problems. Instead it provides a plausibility of (incompatible) answers and a multitude of (different and distinct) solutions. It does not even pretend to offer the

---

39 For Christian Wolff, philosophy is *scientia possibilium, quatenus esse possunt* (*Philosophia Rationalis*, sect. 29).
40 Russell 1959, pp. 249–50.

truth but only surveys of different and discordant *purported* truths emanating from different purporters. Philosophy as such cannot abandon the quest for credible truth regarding the solution of philosophical problems. Non-Hegelian plausibility syncretism does not even attempt to provide it.

Facing a plurality of contending rival answers to philosophical questions, the sceptic embargoes *all* of the available options and enjoins us to reject the whole lot as meaningless or otherwise untenable. A more radical option, though equally egalitarian, is to proceed in the exactly opposite way and view all the alternatives positively, embracing the whole lot of them. The guiding idea of this approach is that of *conjoining* the alternatives. Such a syncretism represents an attempt to "rise above the quarrel" of conflicting doctrines, refusing to "take sides" by taking all the sides at once. It is a Will Rogers kind of pluralism that never met a position it didn't like. Confronted by discordant possibilities, it embraces them all in a generous spirit of liberalism that sees them all as potentially meritorious. But of course what is now at issue is mere plausibility and not actual truth.

The discordant doctrines of philosophers are seen by syncretism as no more than individual contributions to a communal project whose mission is not a matter of *establishing a position* at all but one of *examining positions, of exploring the entire space of alternatives.* The key question is now not (as with the hermeneutic approach described above) the history-oriented "What positions *have been* taken?" but the possibility-oriented "What positions *can be* taken?" It is a matter of the comprehensive appreciation of possibilities in general and not one of trying to substantiate some one particular position as rationally appropriate. On this approach, the real task of philosophy is to inventory the possibilities for human understanding with respect to philosophical issues. In studying the issues we widen our sensibilities, enhance the range of our awareness, and enlarge the range of our cognitive experience. Philosophy now becomes a matter of horizon broadening rather than problem solving-a matter not of *knowledge* at all but of the sort of "wisdom" at issue in an open-endedly welcoming stance toward diverse positions. To take philosophy as *judging* its theses and theories deeming these acceptable and those not looks from this standpoint to be something of a corruption. Instead, philosophy is seen as essentially nonjudgmental, its task being to enlarge our views and extend our intellectual sympathies by keeping the entire range of possibilities before our mind.

But even without taking this line, it is clear that the benefit that a multilateral approach based on plausibility aporetics lies in its enabling us to realize lies in its enabling us to see clearly

- How our own system is related to its rivals;
- How our own system emerges from its antecedents;

- What price we are paying and what benefits we are denied from doing things our way rather than in the way of alternatives;
- Provide ways of processing the actualizes (and disactualizes) of our own position vis-à-vis atheism.

Philosophy-at-Large vs. My Philosophy

In the end, the fact remains that a plausibilistic systematization of philosophy-at-large is seems to offer an invitation to error. For it would be a profound error to regard the result as a system of philosophy and to regard the sort of syncretism involved as a way of or substitution for philosophizing. After all, the quest for truth in relation to its questions is the very reason for being of the enterprise. To philosophize is in the end something that calls for working out our own answers to the questions. A canvas of possible answers—however interesting and indeed useful for philosophizing it may be—is in the end no way of answering the questions.

How, then, should one see *one's own* philosophy in its relation to philosophy-at-large? It is, obviously and of course, a constituent *part* thereof: that goes without saying. But of course one's own attitude toward it is—and should be—very special, specific—and prejudicial. After all, the philosopher's position at issue just would not be *my* philosophy if I did not think it to be correct. In taking a philosophical position of one's own, one is bound to see it as unquestionably correct and the others as erroneous, no matter how plausible they may appear to be. Our cognitive ventures are generally preoccupied with truths or at least what we take to be such. And the manifold of truth—of genuine and authentic truth—has three salient features: it is alternative excluding (admits no self-contradiction), internally coherent and deductively closed. (Already Aristotle, the father of logic, effectively insisted on the matter.)

But of course this does *not* mean that I must see these positions as worthless. In regarding philosophy at large as a matter of plausibility there is, of course, nothing to prevent my regarding my own philosophy as a matter of truth. After all error is not evicted equal—some of it can hit very close to the mark. And it can in principle exhibit details of workmanship. And at the least it can provide us with constructive illustrations of what it takes to avoid systematic pittfalls.

Developing the aporetic conception of philosophy-at-large is not a way of philosophizing. To philosophize we must figure out where *we* stand—we ourselves—and not just elaborate and coordinate the manifold views of philosophy at large. No matter how elaborately and sophisticated we canvas the spectrum of alternatives—of what can be thought—we must in the end make up our minds about what we propose to think on our own account. The analysis of plausibility and its ramifications is all very well in its own way, but it is neither a version of nor a replacement for the pursuit of truth.

Philosophy, after all, is in the business of seeking to answer "the big questions" concerning man and his place in nature's scheme of things. The object of philosophizing is to remove ignorance and puzzlement, to resolve cognitive problems-to provide *information,* in short. To abstain from taking a definite position, to refuse to take sides (be it through sceptical abstention or through open minded conjunctiveness) is simply to abandon the enterprise. To endorse a plurality of answers is to have no answers at all-an unending openness to various possibilities, a constant yes-and-no, leave us in perpetual ignorance.

To be sure, studying the modus operandi of philosophizing will in the end be a *part* of philosophy—who else but a philosopher is willing or able to do such a thing. But this form of metaphilosophy is at most and at best a part of philosophy—it is not and cannot be a substitute for philosophy itself. The study of philosophy at large will not—cannot—itself constitute a philosophy. To survey of what is plausible will not instruct us about what is true. To look into the plausible answers to our questions will not show us what the appropriate answers are.

It is, of course, instructive to explore the manifold of plausibility. But in the end we want more. Plausibility mongering is not enough: we hanker after truth. Shift the realm of deliberation from philosophy at large to my philosophy—to what I can, on my own account, accept as true.

And so, notwithstanding, its obvious attractions, the view of philosophy as possibility exploration is not without its defects. The prime difficulty is that possibility mongering fails to accommodate the central project of serious inquiry. Possibilities don't answer questions. To *engage in* the enterprise, rather than merely to deliberate *about* it, we must ask, "What position *shall* we take?" and not merely "What position *can* we take?" Plausibility mongering is all very well as far as it goes. But in the end if one wants to have actual (rather than merely possible) answers to one's questions, one must make up one's mind. One must stop being a student of philosophy and become a philosopher.

As Bertrand Russell rightly showed in the passage quoted above, we can *study* philosophy to expand our horizons, to learn what sort of positions are available-what sort of stories can be told about the issues. But we *engage* in philosophizing because we want to have answers to the questions and solutions to the problems of the field answers with which we ourselves can rest rationally content, even if they do not form the focus of a universal consensus. If we wish to *philosophize,* to arrive at answers to our questions, we cannot avoid taking a position. We must be committal and espouse views and positions in a selective, discriminating way that endorses some alternatives at the expense of others.

The individual laborer in the domain of philosophy *cannot* transcend his individuality and adopt the diversified labors of the community into his own doctrine. Conjunctionism is a grand notion that founders on the recalcitrant fact that there is just no way of making the library over into a single coherent book by conjoining the multiplicity of discordant philosophies into a meaningful whole. Individual philosophers can do no more than take one stance among others, arriving at a position that unavoidably remains controversial. They cannot at once engage in the enterprise and enjoy the security of a higher vantage point above the din of battle. By their very nature, philosophical disagreements resist being transcended by way of conjunction. Conjunctionism is an invitation to indifference, to refraining from the serious business of making up one's mind.

To project one's pacifist ideology into the sphere of philosophy is to emasculate the subject. It is not by accident that Athena, goddess of wisdom and patroness of philosophy, presides over the arts of war as well. The strife of systems is relentless-the destiny of philosophy is not peace but the sword.

But of course this does not prevent a larger, more generous view of philosophy-at-large. When I contemplate the status of the claims and theses of *my own* philosophical position, I do indeed myself regard them as so many claims to truth. (And just this is why the internal consistency of *my own position* is a critical function). But when I contemplate the work of my philosophical rivals I do not—can not—regard these contentions as truths but I certainly can—and very possibly *should*—see those theses and contentions as *plausible*.

The present deliberations highlight the difference between the *study of philosophy* on the one hand and the *practice of philosophy* on the other. In the study of philosophy a neo-Hegelian plausibility systematization proves to be a highly useful and instructive resource. But of course to practice philosophy we must deal in more than mere plausibilities: to determine where we are to stand we cannot rest content with the examination of plausibilities and possible answers but must decide upon optimalities and appropriate (at least preemptively correct) answers. At this level truth systematization is the best way forword,

To say all this, however is to *distinguish* the two enterprises, not to *detach* them. For of course the quest for answers that are optimally tenable—and thus at least preemptively correct and true—require us to compare the available alternatives. And just what these alternatives are, and what ramifications they send into the wider network of related issues, is just exactly what a plausible systematization can enable us to see.

8 On Validating First Principles

First Principles of Scientific Knowledge: Some Examples

Philosophers have generally understood by a "first principle" (*Grundsatz*) of natural science those principles (*Prinzipien*) regarding "how things work in the world" which (1) relate to the *modus operandi* of nature, and (2) deal with features of it so fundamental and so pervasive that, in a world *lacking* these features, it would be effectively impossible for inquiring creatures such as we men to obtain the information needed to build up natural science as we know it. Such principles pertain to those features of the world which—given inquiring creatures constituted along prevailing lines, and proceeding in the established ways—make possible the achievement of knowledge by the means and methods of natural science.

Principles of this sort relate to the orderliness and to the lawfulness of nature—to its conformity to *rules* of various sorts. If nature were not rulish—if it were "unruly" in these ways—a *scientific* account of the world in terms of lawful aspects would not be possible. This rulishness is basic to the very possibility of natural science. For consider the aims of science, viz., the description, explanation, prediction, and control of nature. It is easily seen that each of these would be altogether unrealizable without rulishness. Rulishness is a precondition for the possibility of the very possibility of science. And indeed the various modes of rulishness at issue in the first principles we have adduced (causality, regularity, coherence uniformity and the rest) are all related to aspects of the workings of nature that underwrite the very possibility of scientific inquiry.

The aim of the present discussion is to examine the nature and function of such first principles, and—above all—to clarify the basis on which their *justification* rests. This question is very much in order, for these principles are clearly not —by their very nature—"self-evident" and in need of no further justification

Rational Substantiation

All too clearly, the first principles from which our inquirers set out cannot be validated with reference to further considerations that are yet more basic. (This is so by hypothesis—if they could be established in this way, they would not be

---

Note: This chapter was originally published in the *Allgemeine Zeitschrift für Philosophie*, vol. 2 (1976), pp. 1–16.

"first" or basic.) Little probative headway can thus be made by trying to provide any sort of "derivation" of these principles by recourse to premises in whose establishment these principles will themselves figure in their characteristic regulative role. And so, since first principles cannot be justified in terms of other, more fundamental premises, they must be justified in terms of their own consequences. Their validation requires a systemic approach. In particular, such principles must be able to *accommodate experience* in smooth attunement to the concrete interactions through which the world's realities make their impact upon us. It must thus be shown that if the principle is rejected then either (1) certain eminently desirable results will be lost, or (2) certain highly negative results will ensue. Accordingly, such principles can only be validated in terms of the unacceptable implications of their abandonment. In sum, first principles are to be judged by how smoothly they fit into the explanatory rationale of our experience with a view in particular to the question of how efficient an instrumentality they provide for the overall explanation and systematization of that experience. The crux here is that our first principles must not only meet the conditions of theoretical systematicity but must do so with reference to experience.

The dialectical process at issue may be clarified in a schematic way as follows. One begins with the presumptive "trial assumption" or "provisional hypothesis" of a certain cognitive mechanism—an instrumentality (process, method) for issue-resolution. One then proceeds to employ this instrumentality so as to determine a body of putative knowledge—an overall system. Thereupon, one deploys this knowledge to provide a rational accommodation for our "experience"—an information at large. Then, one *revises* the initial "trial assumption" (provisional hypotheses) with a view to the successes and failures of these applications. And then starts the process all over again at the first step. What is at issue throughout is not just a merely *retrospective revalidation* in the theoretical order of justification, but an actual *revision or improvement* in the dialectical order of development, a cognitive upgrading of suppositions initially adopted on a tentative basis.

Reflection on this process makes it clear that if *this* is how the first principles of inquiry in question-resolution are legitimated, then the status of such principles is defeasible in the light of "the course of experience"—it becomes *a posteriori* and contingent. This circumstance is one whose importance cannot be overemphasized. It means that no particular formulation of a philosophical position —no explicitly stated substantive resolution to a philosophical problem—can be altogether adequate as it actually stands, without further explanation, qualification, and explanatory exposition. Further questions will always arise that need to be addressed in the larger scheme of things.

Descartes says that only physical things and intelligent beings exist. But what then of animals? Plato maintains that mathematical objects like shapes and numbers exist in a separate realm altogether apart from the material world. But how then can we embodied humans know them? Once a substantive philosophical thesis is formulated, further questions about its meaning, implications, bearing, and purport will always arise. As it stands, in its actual and overt formulation, the thesis is not complete, not quite correct, not altogether adequate to what needs to be said on the subject. Under the pressure of an ongoing readjustment to an ever-widening context of considerations, it admits of various alternative interpretations, constructions, elaborations; it presents further issues that must be resolved; it requires explanation, exposition, qualification. Taken just as it stands, without further elaboration, the exposition is not satisfactory: it leaves loose ends and admits of undermining objections.

In examining our first principles—and thus the philosophical theses that hinge upon them—we accordingly embark on a cyclically repetitive (and thus in theory nonterminating) process of elaboration and subsequent reformulation.

Such a dialectic of contention and elaborative explanation engenders an ever more fine-ground detail the inner commitments and involvements of the initial position that was the starting point of our endeavor to answer the philosophical question at issue. With any substantive philosophical issue, the intellectual game of problem-solving and issue resolution can thus be played at ever more elaborate levels of sophistication.

The ongoing elaboration of a philosophical position constitutes a process of expository development that brings its various aspects into clearer and sharper focus. The continuing development of conceptual machinery provides a process of *ideational* magnification analogous to the process of perceptual magnification that accompanies the ongoing development of the physical machinery of microscopy. And there is no reason of principle why this process of ongoing elaboration and sophistication need ever stop; it can continue as long as our patience and energy and interest hold out. When we stop, it is because we are sufficiently wearied to rest content, and not because the project as such is completed.

It emerges on this perspective that the first principles that are basic to philosophical understanding are "first" (and ultimate questions "ultimate") only in the first instance or in the first analysis and not in the final instance and the final analysis. Their firstness represents but a single "moment" in the larger picture of the dialectic of legitimation. They do not mark the dead-end of a *ne plus ultra* that admits no further elaboration and substantiation. The question "Why these principles rather than something else?" is certainly *not* illegitimate here. It is something we can not only ask but also answer, even if only provisionally and imperfectly, in terms of the complex dialectic afforded by the cyclic structure of legitimation as sketched above.

This approach indicates that philosophy should be seen as being, in the end, erected on a contingent and ultimately factual basis. Its determinative first principles and their correlative substantive doctrinal contentions are seen as defeasible and defensible: they can and need to be legitimated—a process that proceeds in the light of empirical considerations. Forming, as they do, an integral component of the cognitive methods that have evolved over the course of time, it can be said of them—as of other strictly methodological instrumentalities—that "*die Weltgeschichte ist das Weltgericht*," or, loosely translated, "The proof of the pudding is in the eating." (Recall too Hegel's penetrating dictum that metaphysics must follow experience and not precede it.)

These observations go no farther than to say that circumstances *could* arise in which even those very fundamental first principles that define for us the very idea of a philosophical category might have to be given up. But to concede the *possibility* is not—of course—to grant the likelihood—let alone the reality. Once entrenched, the principles at issue are so integral a component of *our* rationality that we ourselves cannot even conceive of *any* rationality that dispenses with them: we can readily conceive *that* they might have to be abandoned, but can scarcely conceive *how*. Thus to concede the in-principle defeasibility of these principles does nothing to undermine their indispensibility for us now, in the present state of the art in our inquiries.

The Function of First Principles: Their Regulative Status

The most basic role of first principles is not to serve a *substantive* (or descriptive) function in describing the nature of the world, but rather toe serve a *regulative* (or methodological) function in providing the testing standards for checking the acceptability of such substantial accounts. They provide criteria of qualification for a "rationally adequate account" or for an "intellectually satisfactory picture" of what goes on in the world. They are *conditions of adequate understanding*—for assessing whether we might conceivably have "got it right" in describing the workings of nature. They are not so much the products of as the preconditions for a rationally adequate account of nature.

This explains the firstness of "first principles" (*die Grundheit der Grundsatze*). They are "first"—that is, prior to *what* we understand of the workings of the world, because they represent defining conditions of the circumstance *that* we actually might understand it aright. They correspond to *qualifying preconditions* for an acceptable theoretical picture of the world. Fundamentally and in the first instance (the need for this qualification will become clear in the subsequent discussion), first principles thus function in a *regulative* role. That is, they regulate and control the claims to rational acceptability of our explanatory-descriptive accounts of the world. They represent a check on the validity of our pretensions to understand how things work in the world. In sum, they serve as *regulative conditions of understanding*

If a characterization of the workings of nature were to violate these principles—if it somehow abrogated the unity or the consistency or the lawfulness of nature—then it would *eo ipso* thereby blazon forth its own inadequacy. One could not rationally rest content with such an account because *(ex hypothesi)* it contravenes what is in fact one of the very characterizing conditions of an adequate account.

Consider, for example, the principle of the uniformity of nature, construed to the effect that if case C is subject to law L and the circumstance of case C' are sufficiently like those of C, then C' must also be explained with reference to L.[41] It is clear on the very surface of it that this is a procedural principle governing the acceptability of explanations. Much the same holds for the principle of the simplicity of nature ("of alternative and in most other respects comparable accounts, accept the simplest"), the principle of the consistency of nature ("mutually inconsistent accounts cannot be accepted"), and the various other principles at issue here.

Thus first principles—considered in this, their fundamental role—(1) are *procedural* rules (i.e., relate to *what to do)*, and (2) serve as *controls* in a qualificatory, normative function in relation to the acceptability of descriptions or explanations.

The Ultimate Justification of First Principles

How are such regulative first principles of empirical knowledge to be validated? This issue is problematic because the seemingly straightforward way of arguing for these principles—viz., that after all they *must obtain,* because if they didn't, it would be impossible to achieve scientific knowledge—leaves itself open to the skeptic's reply that our *vaunted* scientific knowledge may well itself be spurious and self-deluded. Given that science *presupposes* these principles we cannot without circularity argue that it also *justifies* them. The venture of seeking to val-

---

41 The old universalistic construction of this thesis is naive and runs afoul of stochastic processes.

idate first principles by a justificatory recourse to further theses about the ways of the world is unsatisfying and unsatisfactory.

To be sure, one can (and should) give a metaphysical or cosmological "deduction" of these principles that assures their validity on the basis of how things usually work in the world. But it is not a probatively sufficient procedure to seek to derive them from theses about the nature of the world (that nature is one comprehensive whole, that nature is everywhere consistent, etc.). For such a way of proceeding is largely circular. If the decision to accept a certain account of nature is indeed controlled by the criterion of the conformity of this account with such principles as those of unity, consistency, etc.—then there can be no probative advance in scrutinizing such an account, once accepted, and remarking "Lo! Nature is unified (consistent, etc.) ". This way of proceeding would clearly involve a vitiating circularity.

Little probative headway can thus be made by trying to provide any sort of "derivation" of these principles with a recourse to premisses in whose establishment these principles will themselves figure in their characteristic regulative role. At best what we have here is a kind of negative control: something would clearly be amiss if the cosmologico-metaphysical ideas we invoke to justify our inquiry-methods were not themselves revalidated by the results obtaining by putting these methods into operation.

The Constitutive Transmutation of First Principles

First principles are, as we have said, fundamentally and in the first instance *regulative;* their basic task is to provide certain acceptability-conditions for a "rationally adequate account" of what goes on in the world. But this is only how the matter stands *in the first instance.* In the final analysis, these principles will clearly also have a constitutive job to do.

This is readily shown by an illustration, say with respect to the principle of the uniformity of nature.

> *Regulative principle:* Any acceptable account must treat some issue in different settings along uniform lines; explain similar in similar ways. *Constitutive counterpart:* Nature is uniform—whenever something happens in accordance with certain laws in one place (time, context) then something similar will happen in accordance with these selfsame laws in another place (time, context).

It is clear that adhesion to the regulative principle will limit our horizons of acceptability to those (substantive) accounts in which the "constitutive counterpart" to this regulative principle obtains. Accordingly, such adhesion assures that whatever account one in fact accepts will conform to the specifications of the constitutive counterpart of the regulative principle at issue. If *those* principles provide the preconditions of acceptability of descriptive or explanatory accounts then *these* (correlative) factors must be operative within the account.

The result here is a *substantive transmutation* of a fundamentally regulative principle. Such regulative principles accordingly become the locus of *a priori* features of the world. Certain features must inevitably obtain of the world because only those descriptive and explanatory accounts which are consonant with these features are to be regarded as qualified for rational acceptability.

The Intelligibility of Nature

This general point receives a special reinforcement in the present context, thanks to the particularly vital role played by the regulative first principles at issue in setting the very conditions of intelligibility for explanatory descriptions of nature.

In the first instance or approximation (but *only* in the first instance or approximation), a seemingly *ontology oriented* question like" Why is the world lawful" can be answered *epistemologically*. Instead of a *ratio essendi* (causal reason) we can adduce a *ratio dicendi* (reason for claiming). Why is the world lawful? = Why does our account of the world provide for the prevailing operation of natural laws? Because *lawfulness* represents a regulative first principle of scientific cognition. We wouldn't accept an account as "an adequate account of the world" if it didn't have a proper place for laws.

Why is the world uniform? = Why does our account of the world provide for a prevailing uniformity of mechanism and process? Because uniformity is a regulative principle. One wouldn't (be entitled to) accept an account as "an acceptable account of the world" if it did not proceed along uniform lines.

And the same account holds for consistency, coherence, and all the other elements of intelligibility. They function regulatively with respect to rationality and intelligibility.

Suppose (for sake of discussion) not a principle of unity/uniformity of nature, but a principle of diversity/disuniformity of nature (nature never repeats, never does the same thing the same way twice, etc.). Then, of course, adoption of this principle at the regulative level ("Never accept an account which has it that...") means that at the substantive level nature will be disuniform.

But a drastic incongruity would now result. Nature *nolens volens* becomes uniform in her avoidance of uniformity. We are enjoined to avoid uniformity, and yet enjoined to accept uniform results. Such a condition is perhaps not actually self-contradictory, but *incongruous* and dissonant. The theory does not support or substantiate itself.

The very opposite is the case with regard to those regulative principles that we in fact espouse: coherence, consistency, unity, uniformity, etc. They assure a unity of approach, a uniformity between the methodological groundwork of the account and its actual content, a: parallelism of the internal aspects of result and the external aspects of method. The theory substantiates .itself as it is only congruous that it should. Coherence thus plays a special role in self-substantiation; indeed, self-substantiation is simply *an aspect* of coherence/uniformity.

Self-substantiation is accordingly a crucial part of intelligibility. It represents a very general desideratum. A logic should be developed logically on its own showing (autodescriptively). Similarly, a system of nature should be developed systematically—its productive principles should be consonant with its productions, its practice should accord with its teaching. Such selfsubstantiation or functional isomorphism between the tenor of the developing principles and the developed structures to which they lead is clearly a fundamental desideratum. This points to the crucial metatheoretic aspect qualifying of first principles as principles of intelligibility.

The Test of Systematization

The flaws and failings that one can in principle encounter in philosophizing are many. There are flaws of issue-formulation (ignoring issues, distorting issues), in data management (inappropriately bestowing or denying credence, for example by "flying in the face of plain fact" or failing to reckon appropriately with technical or scientific knowledge), in position construction (oversimplifying or overcomplicating one's resolutions of problems relative to the data—in the extreme case by actually contradicting oneself). Then too there are flaws of argumentation (in reasoning for one's conclusions or in refuting counter-argumentation), and flaws in dialectics (in giving too short a shrift to unpopular issues or assaulting one's opponents position only at their weakest spots without reckoning with their strengths). But all such flaws are, in the end, so many procedural failures within the setting of one large project: that of systematization. For the strength and weakness of a philosophical position comes down to the extent to which it is developed systematically in relation to the problems it confronts and to the rival alternative resolutions that it outweighs.

To obtain a clearer view of the underlying rationale for systematization as the instrument of truth-estimative conjecture in philosophy, let us glance back once more to the epistemological role of systematicity in its historical aspect. The historical point of departure here was the classical view that the principle of *adaequatio ad rem* is to be such construed as to mean that since reality is systematic, an adequate account of it will also be so. On this basis, philosophers long saw system as a crucial aspect of truth, stressing overall systematicity of "the truth," and holding that the totality of true theses must constitute a cohesive system. This classical approach viewed systematization as a two-step process: first determine the truths, and then systematize them. (Think of the analogy of building: first assemble the bricks, then build the wall.) With the tradition from Leibniz through Kant to Hegel, however, we come to a reversal of this approach. We now move beyond "*true → systematic*" to embrace the reverse transition "*systematic → true.*" Fit itself affords the criterion of rightness. From a *desideratum of the organization* of our body of (presumed) factual knowledge, systematicity accordingly came to be metamorphosed into *a qualifying test of membership*—a standard of facticity.[42] The effect of this Hegelian inversion of the traditional relationship of systematicity to truth is to establish "the claim of system as an arbiter of fact," to use F. H. Bradley's apt expression.

Beginning with the implication-thesis that what belongs to "our knowledge" can be systematized, we are to transpose it into the converse: If a proposition is smoothly co-systematizable with the whole of our (purported) knowledge, then it should be accepted as a part thereof.

From being a characteristic of "accepted knowledge" (as per the regulative idea that a body of knowledge-claims cannot properly qualify as such if it lacks a systematic articulation), systematicity is now transmuted into a testing standard of (presumptive) truth—an acceptability criterion. The key idea at issue is a transformation of systematicity from a framework for *organizing* knowledge into a mechanism for *determining* adequate knowledge-claims. Fit, attunement, and systematic connection thus become the determinative criteria for assessing the acceptability of claims, the monitors of cognitive adequacy.

This idea of systematicity as an *arbiter* of knowledge was implicit in Hegel himself, and developed by his followers, particularly those of the English Hegelian school inaugurated by T. H. Green. This Hegelian inversion leads to one of the central themes of the present discussion—the idea of using systematization as a control of substantive knowledge. F. H. Bradley put the matter as follows:

> The test [of truth] which I advocate is the idea of a whole of knowledge as wide and as consistent as may be. In speaking of system [as the standard of truth] I always mean the union of these two aspects [of coherence and comprehensiveness] ... [which] are for me insepara-

---

42 Compare Bradley 1914, pp. 202–18.

bly included in the idea of system. ... Facts for it [i.e., my view] are true ... just so far as they contribute to the order of experience. If by taking certain judgements ... as true, I can get some system into my world, then these "facts" are so far true, and if by taking certain "facts" as errors I can order my experience better, than these "facts" are errors.[43]

The plausibility of such an approach is easy to see. Pilate's question is still relevant. How are we humans—imperfect mortals dwelling in this imperfect sublunary sphere—to determine where "the real truth" lies? The Recording Angel does not whisper it into our ears. (If he did, I doubt that we would understand him!) The consideration that we have no *direct* access to the truth regarding the *modus operandi* of the world we inhabit is perhaps the most fundamental fact of epistemology. We must recognize that there is no prospect of assessing the truth—or presumptive truth—of knowledge claims independently of our efforts at systematization in rational inquiry. The Hegelian idea of truth-assessment through systematization represents a determined and inherently attractive effort to adjust and accommodate to this fundamental fact.

On this approach, then—to which the present view of philosophizing as truth-estimative conjecture fully conforms—the assurance of inductively authorized contentions turns exactly on this issue of tightness of fit: of consilience, mutual interconnection, and systemic enmeshment. Systematicity becomes our test of truth, the guiding standard of the truth-estimation at issue in the process of "rational conjecture" that defines philosophizing. Our "picture of the real" is thus taken to emerge as an intellectual product achieved under the control of the idea that systematicity is a regulative principle for our theorizing. Here, evidentiation and systematicity are inextricably correlative.

To be sure, the line of approach at issue here has weakened the move from systematicity to truth somewhat. For its epistemological position moves from systematicity not to correctness itself, but rather to *the rational warrant of claims to correctness*. The operative transition is not from "systematic" to "correct," but rather from "systematic" to "rationally claimed to be correct." The role of systematicity is, in the first instance, epistemic (and only derivatively ontological). The "best available answer" at issue through the systematization of experience is so only in the sense of affording us the best available *estimate* of the truth that we can make in the circumstances—which, being the most we can possibly obtain, is (or should be) sufficient to content us.

---

43 Bradley 1914, pp. 202–218; see pp. 202–203 and 210.

Rhetoric vs. Demonstration: The Complexity of Systemic Justification

The fact is that philosophy cannot provide a rational explanation for *everything*, rationalizing all of its claims "all the way down." Sooner or later the process of explanation and rationalization must—to all appearances—come to a halt in the acceptance of unexplained explainers. David Hume, for example, wrote that the power of the imagination in collecting ideas and presenting them to consciousness is directed "by a kind of magical faculty in the soul, which, tho' it be most perfect in the greatest geniuses, and is properly what we call a genius, is however inexplicable by the utmost efforts of human understanding."[44] And in a similar view, Immanuel Kant proclaimed "how in a thinking subject *outer intuition*, namely that of space, with its filling in of shape and motion, is possible ... is a question which no man can possibly answer" (*C. Pu. R.*, A 393). And again "the question ... of how in general a communion of [interacting] substances is possible ... is a question which ... [one] will not hesitate to regard as likewise lying outside the field of all human knowledge" (*ibid.*, B 428). Among 17th century philosophers (Descartes, Berkely, Leibniz) such insolubilia were standardly laid at the door of God. Nor have the philosophers of a more recent time given up on characterizing certain facts as inexplicable.[45]

Now if we think of explanation as proceeding linearly, in the manner of logical derivation, by explaining A in terms of B which is in its turn explained in terms of C, and this in turn referred to D, then of course we must accept some inexplicable ultimate—unless we are to descend into an infinite regress. But if we are prepared to think of explanation as holistically systemic, then we can explain each of the group A, B, C, D, in terms of its being optimally attuned to all the rest in their collective wider context. And this means that we must accept some contentions *prospectively*—not because they rest on solid established grounds but because they lead to promising results—because they have a certain systemic plausibility about them within the overall setting in which they figure.[46]

This fact that philosophical exposition cannot in the end operate satisfactorily in the linear manner of an axiom system proceeding from a starter-set of self-evident truths is reflected in the nature of philosophical exposition. There are two very different modes of writing philosophy. The one pivots on inferential ex-

---

[44] Hume 1967, p. 24.
[45] The scientific intelligibility of nature, for example. See the discussion of this issue in Rescher 1987.
[46] This coherentist aspect of philosophical methodology will be dealt with in more detail in the next chapter.

pressions such as "because," "since," "therefore," "has the consequence that," "and so cannot," "must accordingly," and the like. The other bristles with adjectives of approbation or derogation—"evident," "sensible," "untenable," "absurd," "inappropriate," "unscientific," and comparable adverbs like "evidently," "obviously," "foolishly," etc. The former relies primarily on inference and argumentation to substantiate its claims, the latter primarily on the rhetoric of persuasion. The one seeks to secure the reader's (or auditor's) assent by reasoning, the other by an appeal to values and appraisals—and above all by an appeal to fittingness and consonance within the overall scheme of things. The one looks foundationally towards secure certainties, the other coherentially towards systemic fit with infirm but nevertheless respectable plausiblities. Like inferential reasoning, rhetoric too is a venture of justificatory systematization, albeit one of a rather different kind.

Consider the following passage from Nietzsche's *Geneaology of Morals* (with characterizations of approbation/derogation indicated by being italicized):

> It is in the sphere of contracts and legal obligations that the moral universe of guilt, conscience, and duty, (*"sacred" duty*) took its inception. Those beginnings were *liberally sprinkled with blood*, as are the beginnings of *everything great on earth*. (And may we not say that ethics has never lost its *reek of blood* and torture—not even in Kant, whose categorical imperative *smacks of cruelty*?) It was then that the *sinister knitting together* of the two ideas guilt and pain first occurred, which by now have become quite inextricable. Let us ask once more: in what sense could pain constitute repayment of a debt? In the sense that to make someone suffer was *a supreme pleasure*. To behold suffering gives pleasure, but to cause another to suffer affords an *even greater pleasure*. This *severe statement* expresses an old, powerful, *human, all too human sentiment*—though the monkeys too might endorse it, for it is reported that they heralded and preluded man in the devising of *bizarre cruelties*. There is no feast without cruelty, as man's entire history attests. Punishment, too, has its *festive features*.[47]

Not only is the passage replete with devices of evaluative (i.e. positive/negative) characterizations, but observe too the total absence of inferential expressions. We are, clearly, *invited* to draw certain unstated evaluative conclusions. But the inference, "Man is by *nature* given to cruelty, and therefore cruelty—being a natural and congenial tendency of ours—is not something bad, something deserving condemnation," is left wholly implicit. This conclusion at which the discussion aims is hinted at but never stated, implied but never maintained. In consequence, reason can gain no fulcrum for pressing the plausible objection: "And why should something natural automatically be therefore good: why should the

---

[47] Nietzsche 1923, Essay II, Sect. 6.

primitiveness of a sentiment or mode of behavior safeguard it against a negative evaluation?" By leaving the reader to his own conclusion-drawing devices, Nietzsche relieves himself of the labor of argumentation. Not troubling to formulate his position, he feels no need to give it *support*; he is quite content to *insinuate* it.

By contrast to the preceding Nietzsche passage, consider the following ideologically kindred passage from Hume's *Treatise* (with evaluative terms italicized and inferential terms capitalized):

> Now, SINCE the distinguishing impressions by which moral good or evil is known are nothing but particular pains or pleasures, IT FOLLOWS that in all inquiries concerning these moral distinctions IT WILL BE SUFFICIENT TO SHOW the principles which make us feel a satisfaction or uneasiness from the survey of any character, IN ORDER TO SATISFY US WHY the character is *laudable* or *blamable*. An action, or sentiment, or character, is *virtuous* or *vicious*; WHY? BECAUSE its view causes a pleasure or uneasiness of a particular kind. In giving a reason, THEREFORE, for the pleasure or uneasiness, we sufficiently explain the vice or virtue. To have the sense of virtue is nothing but to feel a satisfaction of a particular kind from the contemplation of a character. The very feeling constitutes our praise or admiration. We go no further; nor do we inquire into the cause of the satisfaction. WE DO NOT INFER a character to be *virtuous* BECAUSE it pleases; but in feeling that it pleases after such a particular manner we in effect feel that it is *virtuous*. The case is the same as in our judgments concerning all kinds of beauty, and tastes, and sensations. Our approbation is IMPLIED in the immediate pleasure they convey to us.[48]

While for Nietzsche cruelty is effectively a virtue only because people are held to be generally pleased by engaging in its practice, for Hume it is something negative only in that people are generally displeased by witnessing it. The positions differ but their ideological kinship is clear; both writers agree that cruelty is not something that is inherently bad as such—for them the pro- or con-reaction by people is all-determinative.

But what is also clear is that these kindred positions are advanced in very different ways. In the Nietzsche passage, the "argumentation ratio" of inferential to evaluative expressions is 0:12, in the Hume passage it is 9:6. Hume, in effect, seeks to *reason* his readers into agreement by a deduction from "plain facts"; Nietzsche seeks to *coax* them into it by an appeal to conceded suppositions and prejudgments.

Reflection on the contrast between the argumentative and the rhetorical modes of philosophical exposition leads to the realization that these two styles are congenial to rather different objectives. The demonstrative/argumentative

---

[48] Hume 1911, Bk. III, Pt. I, Sect. 2.

(inferential) mode is efficient for securing a reader's assent to certain claims, to influencing one's *beliefs*. The rhetorical (evocative) mode is optimal for inducing a reader to adopt certain preferences, to shaping or influencing one's *priorities and evaluations*.

The *apodictic* (argumentative or probative) mode of philosophical exposition is by nature geared to enlisting the reader's assent to certain theses or theories. It is coordinated to a view of philosophy that sees the discipline in *information-oriented* terms, as preoccupied with the answering of certain questions: the solution of certain cognitive problems. It aims primarily to *convince* by way of reasoning.

By contrast, the rhetorical, *prohairetic* (preferential or evocative) mode of philosophical exposition is by nature geared to securing acceptance with respect to *evaluations:* to enlisting the reader's agreement to certain priorities or appraisals. It is preoccupied with evaluation, with forming—or reforming—our sensibilities with respect to the *value* and, above all, the *importance* of various items. It is bound up with a view of philosophy that sees the discipline in *axiological* terms, as an enterprise that has as its prime task the securing of certain evaluative determinations and the establishment of certain prizings and priorities. It aims primarily to *induce* people to an evaluative standpoint. It exerts its appeal not in reasoning from prior philosophical givens, but rather by rhetorical means that exert their impetus *directly* upon the cognitive values and sympathies that we have fixed on the basis of our experience of the world's ways.[49]

To exert rational pressure on a reader's sensibilities without using arguments that are themselves already value-invoking, one must appeal to the persuasive impetus of this person's body of experiences. Here too providing information can help—but only by way of influencing the sensibility, the reader's way of looking at things and appraising them through an appeal to things one already knows perfectly well. Accordingly, it is here that the rhetorical method comes into its own by enabling an exposition to appeal to—and if need be influence and modify—a reader's body of experiences in order to induce a suitable adjustment of evaluations. There are, of course, many ways to realize this sort of objective. A collection of suitably constituted illustrations and examples, a survey of selected historical episodes that serve as instructive case studies ("History teaching by examples"), or a vividly articulated fiction can all orient a reader's evaluative sentiments in a chosen direction—as the philosophical methodology of Ludwig Wittgenstein amply illustrates. And so, of course, can pure invective, if sufficiently clever in its articulation. What matters is that our agreement is elicited via the fact that something is rendered plausible and ac-

---

[49] Compare Johnstone 1959.

ceptable through its consonance with duly highlighted aspects of our experience —so that the course of our experience itself invites and elicits acceptance.

Since two distinct views of the mission of the enterprise are at issue with the demonstrative and evocative approaches to philosophy, any debate over the respective merits of the two modes of philosophical exposition is thereby inseparable from a dispute about the nature of philosophy. The quarrel is ultimately one of ownership: to whom does the discipline of philosophy properly belong, to the argumentative demonstrators or to the evocative rhetoricians?

This contest over the ownership of philosophy has been going on since the very inception of the subject. Among the Presocratics, the Milesians founded a "nature philosophy" addressed primarily at issues we should nowadays classify as scientific, while such thinkers as Xenophanes, Heraclitus, and Pythagoras took an evaluative—evocative and "literary"—approach to philosophy, illustrated by the following dictum which Kirk and Raven affiliate with Pythagoras:

> Life is like a festival; just as some come to the festival to compete, some to ply their trade, but the best people come as spectators, so in life slavish men go hunting for fame or gain, the philosophers for the truth.[50]

In 19th Century Germany philosophy, Hegel and his school typified the scientific/demonstrative approach, while the "post-moderns" who were their opponents—Schopenhauer, Kierkegaard, Nietzsche—all exemplify the axiological/rhetorical approach. In the 20th century, the scientistic movement represented by logical positivism vociferously insisted on using the methodology of demonstration, while their anti-rationalistic opponents among the existentialists and also among the neo-Romantic theoreticians of Spain (preeminently including Unamuno and Ortega y Gasset) restored extensively to predominantly literary devices to promulgate their views—to such an extent that their demonstration-minded opponents sought to exile their work from philosophy into literature, journalism, or some such less "serious" mode of intellectual endeavor.

In this connection we see as clearly as anywhere else the tendency among philosophers towards defining the entire subject in such a way that their own type of work is central to the enterprise and that their own favored methodology becomes definitive for the way in which work in the field should properly be done. The absence of that urbanity which enables one to see other people's ways of doing things as appropriate and (in *their* circumstances) acceptable is thus perhaps the most widespread and characteristic failing of practicing philosophers. But the fact remains that while individual *philosophers* generally have no

---

50 Kirk and Raven 1957, frag. 278.

alternative but to choose one particular mode of philosophizing as focus of their allegiance, *philosophy* as such has to accommodate both. Philosophy as such is broader than any one philosopher's philosophy.

The irony of the situation is that philosophers simply cannot dispense altogether with the methodology they affect to reject and despise. Even the most demonstration-minded philosopher cannot avoid entanglement in evaluation by rhetorical devices. For even the most rationalistic of thinkers cannot argue demonstratively for everything, "all the way down," so to speak. At some point a philosopher must invite assent through an appeal to sympathetic acquiescence based on experience as such. On the other hand, even the most sentimental philosopher cannot altogether avert argumentation. For a reliance on certain *standards* of assessment is inescapably present in those proffered evaluations, and this issue of appropriateness cannot be addressed satisfactorily without some recourse to reasons. Ironically then, the two modes of philosophy are locked into an uneasy but indissoluble union. While neither the logical (demonstrative) nor the rhetorical (evocative) school can feel altogether comfortable about using the methodology favored by the other, it lies in the rational structure of the situation that neither side can manage altogether to get on without it. The practice of philosophy is ultimately a matter of striving for a smooth systemic closure between the projections of reason and the data of experience—a harmonization in which the two modes of philosophizing come into mutually supportive overall harmonization.[51]

The Need for Quality-Control

This sort of self-substantiation is necessary but is clearly not sufficient: it is an indispensable requisite, but it does not provide the external quality-control of a sufficiently rigorous reality-principle. For such an external quality control is clearly necessary to lay totally to rest the charge of vitiating circularity (as already mooted above).

The legitimation of the methodological tools of inquiry lies *in justifying the whole framework of "ordinary inquiry*—the entire methodological modus *operandi* of which the tools at issue are a part. The legitimation of the methodology of inquiry will naturally carry along with it that of the various components and constituents of the methodology.

---

[51] Rescher 1976b.

But how is an entire framework of inquiry to be legitimated? Clearly by its results—by that dialectical feedback process that validates the workings of the method in terms of its products.

The most promising course is thus *to approach the issue from a methodology oriented point of departure*. Given the regulative, procedural, in short, *methodological* character of first principles, the question can be handled in the standard way by which a method is justified—namely by the *pragmatic* route of the questions "Does it work?" "Is it successful in conducing to the realization of its correlative objectives? "This is patently the right approach to the legitimation of tools, instrumentalities and all other sorts of methodological devices. It seems also to be the right way to tackle the issue of first principles.

The fact is that their ultimate justification is not discursive at all—it does not proceed by the *theoretical* route of a derivation from supporting factual premisses. Rather, their foundation is *pragmatic*, for their legitimation resides in the fact that we men can function effectively in the world, acting on the "knowledge" generated in the context of these regulative principles. Their legitimacy thus does not derive from a theoretical sphere of the *applicative success* of implementing (basing our actions upon) the world-picture built up on this basis. Not a theoretical demonstration of correctness but the sheer element of pragmatic efficacy is the controlling factor of the legitimation of these first principles.

The structure of the over-all justificatory a figure-eight process of legitimation involves the closing of two cycles: Cycle I, the theoretical-intellectual cycle of *consistency between* regulative first principles and their substantive counterparts, and Cycle II, the practical-applicative cycle of *pragmatic efficacy* in implementing the substantive results of the first principles. This second cycle renders the whole framework of first principles vulnerable and subject to defeat in the light of "recalcitrant experience" arising in applications. Its operation means that first principles have a basis that is only seemingly *a priori*, but ultimately *a posteriori* and controlled by a theory external "reality principle".

To be sure—as we shall shortly argue—even all this is only the first stage in a more complex story. For not only our praxis, but its principles as well, change over the course of time in an evolutionary process that is crucial to the legitimation at issue. And this has the consequence of marking first principles not just as an evolutionary product of rationally guided trial and error—an aspect that is crucial to their legitimation in a broader perspective.

The Ramification of Coherence: The Wigner Paradox

The preceding discussion has maintained that—since first principles (such as those of coherence, consistency, uniformity, simplicity) are among the regulative principles of understanding (or rational intelligibility) on the basis of which we judge the acceptability of descriptive and explanatory accounts of what goes on in the world—relative to those principles, it can be guaranteed *a priori* that a world-picture constructed under such a constraint must be such that these factors are descriptively (substantively) present. The *regulatively* controlling role of such principles assures that they must hold in a *substantively* transformed guise as well.

But if this virtually circular process is the basis on which substantive first principles rest, are they not then utterly necessary and altogether impregnable? If we obtain them by way of the regulative-to-substantive route of derivation, does this not make them indefeasible at the substantive level? The answer is negative—simply because our regulative principles are themselves subject to change.

This fact can be made to emerge from the analysis of one typical case. As such, let us consider the principle of the coherence of nature.

Coherence is not so much a single concept as a crossroads or confluence of related concepts. Primarily three major component elements are included here as factors that must be present in an adequate account of nature.
1. Consistency, consonance (lack of discord or dissonance). Not saying in one breath something that is denied in another;
2. Cohesion, connectedness, interrelationship (lack of fragmentation). Not viewing distinct elements as altogether disparate and unrelated;
3. Uniformity, a striving for generality. Seeking to subsume different cases under a common principle and to apply a uniform approach to comparable cases.

Common to these various aspects of coherence is the element of *unity* of presentation and treatment. But if coherence/consistence/uniformity is our "regulative ideal", does it not follow that our world picture must then *inevitably* be coherent? Is the substantive counterpart of a regulative principle not simply an *a priori* but also thereby a necessary truth? The answer to this question is—as we shall now argue—a negative one.

Despite their apriority, it would seem that regulative first principles may actually have to be given up. It is of interest in this context to contemplate one aspect of *incomplete* knowledge that is seldom recognized outside the Hegelian school, namely that one may well have to pay a price for *incompleteness* in terms of *inconsistency*.

After all, if the branches of our science have a limited, incomplete perspective (say, one able to explore Region A, the other Region B), the resulting theories arrived at by extrapolation from an incomplete basis may well be incompatible with one another.

Now a situation of this structural sort seems in fact to be developing in natural science a situation which Eugene P. Wigner (Nobel laureate in physics for 1960) has detailed in the following terms:

> We now have, in physics, two theories of great power and interest: the theory of quantum phenomena and the theory of relativity. These two theories have their roots in mutually exclusive groups of phenomena. Relativity theory applies to macroscopic bodies, such as stars. The event of coincidence, that is in ultimate analysis of collision, is the primitive event in the theory of relativity and defines a point in space-time, or at least would define a point if the colliding particles were infinitely small. Quantum theory has its roots in the microscopic world and, from its point of view, the event of coincidence, or of collision, even if it takes place between particles of no spatial extent, is not primitive and not at all sharply isolated in space-time. The two theories operate with different mathematical concepts—the four dimensional Riemann space and the infinite dimensional Hilbert space, respectively, so far, the two theories could not be united, that is, no mathematical formulation exists to which both of these theories are approximations. All physicists believe that a union of the two theories is inherently possible and that we shall find it. Nevertheless, it is possible also to imagine that no union of the two theories can be found.[52]

There is, of course, no paradox here: it is, of course, perfectly conceivable both that "a union of the, two theories is inherently possible" and that "no union of the two theories can [ever] be found [by us]". For this exact combination can readily arise of the information needed to forge the unifying theory lies "beyond the data barrier" (as we may figuratively put it)—that is; if *ignorabimus* issues are involved in a suitable way. It is important (and perhaps shocking) to realize that incompleteness may well exact its price not simply in *ignorance*—that is, in blanks in our knowledge—but in *inconsistency*. As long as it remains—as it always must—inherently incomplete, we must reckon with a potential infeasibility of imparting to our knowledge that systematic unity and coherence which has been a regulative ideal for science the days of the *episteme* of the Greek philosophers.

---

52 Wigner 1960, pp. 11–12.

It is a very real prospect that science might evolve (in a seemingly settled way) into a Wigner-condition (W-condition) of internal inconsistency. One cannot rule this out on any simply *methodological* grounds of regulative principle.[53]

This line of thought seems to indicate the possibility that even if we pursue our inquiry into nature on the basis of a certain regulative ideal such as coherence, we might nevertheless find ourselves driven to results that to all appearances violate it. Despite all our best attempts to produce (say) a coherent picture of nature, we might nevertheless find ourselves forced *nolens volens* into a position that contravenes it.

But it is clear from the very nature of the case that we are never constrained to accept this sort *of* inconsistency as final and as demanding an ultimate and unavoidable sacrifice of the regulative principle at issue. For, as the very nature of the example indicates, the inconsistency at issue can be viewed as the mere result of *incompleteness*. This train of ideas' shows that even if the prospect mooted by Wigner were realized—even if to all scientific intents and purposes nature simply appears to be inconsistent still, this would not finally refute the principle of the coherence of nature. We are not *forced* to give up our regulative adherence to the principle. For one can (and doubtless would) maintain a commitment to the coherence of nature as a regulative principle. And as a natural consequence of this commitment, one need not deny the possible realization of the W-condition, but one simply interprets this in a certain way, namely as showing that science is incomplete, maintaining that if it were ever rendered more complete (a condition whose realization may conceivably lie beyond our feeble powers) that the anomaly characteristic of the W-condition would be removed.

But consider the following objection:

> Yes this could be done. But is it *rational?* Is it more than "an act of pure faith "—of what is, in the final analysis, a matter of quixotically holding aloft the standard after the troops have been routed from the field?

This objection is in the final analysis decisive. We *can* always save our first principles, but beyond a certain point it becomes *unreasonable* to do so.

---

53 The situation is reminiscent of the late 19th century split between physicists (especially W. Thompson, later Lord Kelvin) on the one hand and geologists and biologists (especially T. H. Huxley on the other) over the issue of the age of the earth. See the discussion in Brush 1969.

The Defeasibility of First Principles

Still, is it really *possible* to give up the ideal of the coherence of nature as a regulative principle? Can this not be ruled out *a priori* by regulative fiat, as altogether conflicting with the very nature of first principles?

A negative answer is in order. It would seem that first principles—even so crucial and significant a first principle as that of the consistency of nature—might indeed be given up (as Boltzmann, for one, has daringly suggested). But how could this be? Do not first principles purport to be necessary and irrefutable rather than contingent and defeasible? Do they not reflect (at the substantive level of the claim that the world *must* have certain features such as coherence and consistency) our regulative determination to sustain these features?

The answer is only a qualified *yes*. They do indeed reflect such a determination-to-sustain with respect to certain principles, but *not to sustain them at all costs*.

First principles can—and should—be legitimated. But they must—to be sure—be legitimated in a very special way, for they clearly cannot be validated by deduction from yet further principles. As such, these principles must be legitimated by a very different sort of move—the sort of move appropriate to methodological instrumentalities in general. The legitimation of the methodological tools *or* inquiry lies *in justifying the whole framework of* "*ordinary inquiry*"—the entire methodological *modus operandi* of which the tool at issue is a part. The legitimation of the methodology of inquiry will naturally carry along in its wake that of the various components and constituents of this methodology

But how is one to legitimate an entire regulative framework of inquiry? Our answer here can be graphically schematized in terms of a dialectical feedback process that validates the workings of the method in terms of its products.

What is thus at issue in the legitimation of first principles is an indirect (oblique) justification with reference to cognitive methodology that is legitimated pragmatically—in terms of its "working out"—its capacity to conduce efficiently and effectively toward the realization of its correlative objectives of satisfactory prediction and control over nature. And the legitimating process at issue here is not a matter of a static circular pattern in which the movement around the cycles occurs in a merely logical rather than temporal order. It reflects an historic process of successive cyclic iterations where all the component elements become more attuned to one another and pressed into smoother mutual conformation.

But this sort of justification proceeds through *experience*, through trial and check; it is empirical, and contingent. This point cannot be overemphasized. The "first principles" on which our factual knowledge of nature hinges are themselves ultimately of an *a posteriori* and factual standing. Their validation is not a matter of abstract theoretical principle but of experience.[54]

The result of this evolutionary process (with the readjustment of first principles as a source of "variation" and pragmatic test as a source of "selection ") is—ultimately—to make the adequacy of our first principles and the success of their functioning seem only natural and "inevitable ". But their aprioricity and necessity is in fact a conditional one—it has no absolute and *theoretical* guarantee at all but is conditional because founded on an ultimately actual basis—that of the *pragmatic* adequacy of successful praxis.

The Contingency of First Principles

But it is clear that if *this* is to be the line of legitimation of the first principles of scientific cognition, then the status of such a principle is defeasible in "the light of "the facts of experience"—it becomes *a posteriori* and contingent.

To be sure, considered as an element *within* the framework of our cognitive commitments, a first principle serves an *a priori* function as regulative guide to our introduction to further elements into the framework. From this *internal* perspective, the principle remains secure: the other elements of the framework are made to give way to it and thus cannot loosen its secure anchoring. If something goes wrong at the theory-internal level of cycle I, then it will be the factual theses that are readjusted and made to give way to the first principles. But if things go amiss at the stage of cycle II of the pragmatic controls that are external to the theoretical cycle I and determine the adequacy of its workings, then the first principles themselves will have to be revised and readjusted. This over-all position is the result of combining two theses: (1) that first principles are, in effect, components of *cognitive methodology,* and (2) that methods are legitimated in terms of praxis—i.e., of applicative success.

Thus first principles are "necessary" from one specific angle of consideration only—namely *internally* to the theoretical framework within which they function in a role that is *a priori* and regulative. But from a wider perspective, one *that introduces the further issue of the legitimation of this entire framework* (an issue, to be sure, that presses outside the cycle of purely theoretical considerations into the area of praxis and pragmatic implementation), we face a situation in which the first principles themselves become subject to abandonment or revision. They have no claims to "necessity" at this stage. Conceivably things

---

54 Compare Rescher 1976.

might have eventuated differently—even as concerns the seemingly *a priori* "first principles of our knowledge". (Why didn't they so eventuate? We can only answer this by the metaphysical component of the "great cycle" of validation.)

First principles are thus "first" and ultimate questions "ultimate" only in the first instance or in the first analysis and not in the final instance and the final analysis. Their theory-internal absoluteness is deceptive—it represents but a single "moment" in the larger picture of the historical dialectic of evolutionary legitimation. They do not mark the dead-end *cul-de-sac* of a *ne plus ultra*. The question "Why these principles rather than something else"? is *not* illegitimate here—it is indeed answerable, even if only complexly so, in terms of the complex dialectic afforded by the cyclic structure of legitimation as described above. This approach yields a meta physic erected on a contingent and ultimately factual basis. The regulative principles and their correlative substantive metaphysical principles are seen as defeasible and defensible: they can and need to be legitimated—a process that proceeds in the light of empirical considerations. (Recall Hegel's thesis that metaphysics must follow experience and not precede it.)

To be sure, all this is to say no more than that circumstances *could* arise in which even those very fundamental first principles that define for us the very idea of the intelligibility of nature might have to be given up. But to concede the *possibility* is not—of course—to grant the likelihood—let alone the reality. The principles at issue are so integral a component of *our* rationality that we cannot even conceive of *any* rationality that dispenses with them: we can conceive *that* they might have to be abandoned, but not *how*.[55]

The very circumstance that these principles are *in theory* vulnerable is the source of their strength *in fact*. They have been tried in the history of science—tried long, hard, and often—and yet not found wanting. They are founded on a solid basis not of arbitrary convention or logical legerdemain, but of trial in the harsh court of experietial reality. Forming, as they do, an integral component of the cognitive methods that have evo.1ved over the course of time can be said that for them—as for other strictly methodological instrumentalities— *"die Weltgeschichte ist das Weltgericht."*

---

55 On this fact-ladeness of the fundamental ideas by whim our very conception of nature is itself framed see Chapter VI, "A Critique of Pure Analysis" of Rescher 1973b.

9 Our View of Reality

Belief and Reality

We have no access to realty (R) save via our thought about it. "Don't bother telling me what you (merely) *think* to be true, but rather just tell me what actually *is* so" is an unmeetable request. We cannot detach our *conception* of reality from our *belief-system* with regard to it. To ask for more is to demand the impossible.

Beliefs must of course be *someone's* beliefs. And so must belief systems. And here there are three prime possibilities:

- B_1 *The Personal: My* belief system;
- B_2 *The Communal: Our* belief system and in particular that of the scientific community of the day;
- B_3 *The Eventual:* The (hypothetically ultimate) belief system of the eventual future's scientific community.

And corresponding to any such belief system, there is the correlative view of reality: reality as that belief system takes it to be. So corresponding to the above trio we have

- R_1 Reality as I myself take it to be;
- R_2 Reality as the scientific community of the day takes it to be;
- R_3 Reality as the scientific community of the eventual future takes it to be [B_3 was C. S. Peirce's favored candidate for Reality itself.]

There is no reason to think that these R_i cannot agree in various particulars, both with one another nor yet with Reality as such. And when this is so that some of my beliefs about the real are generally accepted and will ultimately seen as correct. All this is where things go well. But there is, of course, ample reason to think that those reality-views will differ from one another in various respects.

Moreover if B is take to qualify as the true system of beliefs about reality—the manifold encompassing all correct, true, and accurate facts about it—then there is no cogent reason for thinking that any of those relativized belief systems B_i manages to realize this. On the contrary, we have good reason to believe that $B_i \neq B$. And correspondingly we have every reason to think that any contextualized view of reality realizes the correlative reality correctly. We have every reason to believe that $R_i \neq R_1$.

Scepticism can take many forms, in particular the following:

R_i differs from R in <some, many, every> regard.

With every i there is thus a realistic, a moderate, and an extreme version of skeptical doctrine. It is sensible to the verge of triviality to accept a realistic scepticism at the R_1 level. It is well within the range of plausibility to be moderately skeptical at the R_2 level. But it is well beyond the limits of absurdity to be radically skeptical even merely at the R_1 level.

But just how is it that R_i can fall short of R?

We realize full well that realty differs from our view of it: that we are imperfect knowers with regard to reality-as-a-whole. In specific we are involved in the entire range of errors surveyed in in Display 10.1.

The decisive difference between R and R_i—and the failure of the R_i to do full justice to R—inhere in the fact that at every R_i level none of these cognitive shortcomings can be rules out as implausible, let alone impossible.

Modes of Error

There is no question that the R_i in one way or another afford our best attempts to secure a cognitive grasp on reality. Yet we have no choice but to acknowledge that they fall short in this regard: that we must acknowledge crucial disparities between reality and our knowledge of it.

Display 9.1: Some Cognitive Errors

- Errors of Commission
  $(\forall i)(\exists p)(p \in R \,\&\, p \in R_i)$
- Errors of Omission
  $(\forall i)(\exists p)(p \in R \,\&\, p \notin R_i)$
- Cognitive Incompleteness
  $(\forall i)(\exists p)(p \notin R_i \,\&\, \sim p \notin R_i)$
- Cognitive Indecision (Vagueness)
  $(\forall i)(\exists p)\,(\exists q)[p \lor q] \in R \,\&\, p \notin R_i \,\&\, p \notin R_i)$

And yet these acknowledged imperfections are things we cannot purport and identify. The challenge "Give me an example of an instance where your best available judgment of the truth of the matter is assuredly false" is once more a challenge that we cannot easily meet. Granted, I do and can without obstacle accept the fact of my cognitive imperfection in acknowledging this at the level of generality:

$A(\exists p)(\sim p \,\&\, Ap)$, where A represents acceptance

But I do not and cannot possibly particularize this via:

$(\exists p)A(\sim p \,\&\, Ap)$

For unlike that first thesis, this second entails

$(\exists p)(A{\sim}p \ \& \ Ap)$

And this, being selfcontradictory is clearly absurd.

As regards philosophy in particular, the relation we can envision to obtain between belief and fact is bound to be complicated. And the following circumstances stand at the forefront here.

Lessons

We cannot but accept our own cognitive imperfection. And going beyond this, one must also acknowledge the unrealizability of the requisites of an ideal science that identifies realty with one view of it. The idea of a realized body of knowledge that is altogether complete and correct is an idealization that must be rejected as illusory. Our only cognitive access to the real can at most and at best be bad via an admittedly imperfect approximation to it.

But of course our knowledge of Reality's phenomena is problematic and its shortcomings are undeniable: There are questions we cannot answer, explanations we cannot give, descriptions we cannot provide in detail. But exactly what our cognitive defect are—and specifically where and how our current beliefs are incorrect—are matters about which we are inevitably in the dark. And this itself is a situation that does—or should—enjoin us to a seemly modesty with regard to our philosophical convictions.

The Epistemic Situation of Philosophical Inquiry

Philosophy as a venture in rational deliberation is subject to the following line of thought, also pertinent to human inquiry at large:
1. We confront questions to which we want and indeed need answers, and yet cannot realize them with assured certitude;
2. To secure an answer we have to make the best-available judgment, the best estimate of the correct answer to such questions. [We can do no better than this—our best];
3. Our best estimates and judgement have to be coordinate with (and indeed be determined by) the relevant data at our disposal, while nevertheless;
4. The data at our disposal are both incomplete and instable—i.e. changing in the light of changing conditions and circumstances.

5. Accordingly, we have to acknowledge the venerability tentativably, and potential fragility of our answers and problem-solution.
6. Moreover, we have to acknowledge that in providing those resolutions and answers even the best that we can do may not be good enough, though it would be impracticable and unavailing to ask for more in the prevailing circumstances.
7. In this matter of experience-based question resolution we face the situation depicted in the diagram of Figure 9.1.
8. Clearly the more conjectural a field of inquiry is the less bases we have for warranted assurance in its speculative issue-resolutions.
9. Throughout speculation inquiries there is a fundamental choice that we ourselves can and indeed must make, to wit: How high a degree of warranted assurance do we propose to insist upon before we keep or give answers adequate and acceptable? (And this is an issue we must address with care, for the higher we set the bar, the fewer the range of issues we will be able to answer.)
10. There simply is no absolutely proper and correct way of setting the threshold of evidential assurance. This is not a *factual* issue of determinable correct proceeding: but *practical* issue that calls for a decision rather than a finding.
11. The acceptability of contentions in a field of speculative inquiry will thus in the end fall to practical (i.e. purpose-based reasoning rather than factual observation-based reasoning).

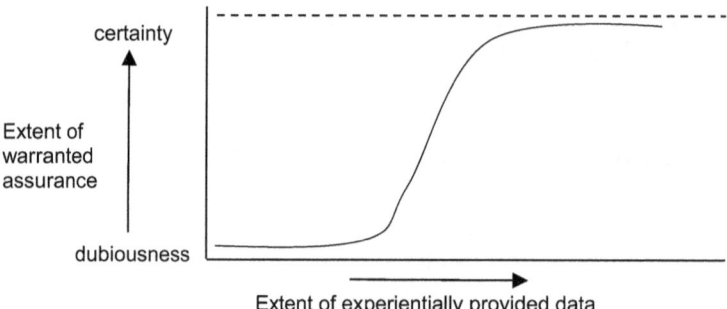

Figure 9.1: Rational Acceptability in the Light of Experience-provided Data. The S-shaped course reflective the relation between experiential value and experiential warranted in matters of rational truth-estimation.

On this basis, our view of reality is—and is bound to be—the product of rational systematization in a manner both filling in detail and comprehensive in scope.

10 Optimalism and Explanatory Totalization

Optimalism

Optimalism is deeply rationalistic. It sees the optimal as being that for whose realization there is the best of reasons—that which would be actual if rationality had it is own way.

What sorts of considerations can possibly provide for the justificatory validation of optimalism? Why should it be that the Principle of Optimality obtains? Why should what is for the best be actual? But consider! To ask this question is to ask for a reason. It is *already* to presume or presuppose the rationality of things, taking the stance that what is so is and must be so for a reason. Once one poses the question "Why should it be that nature has the feature *F?*" it is already too late to raise the issue of nature's rationality. In advancing that question the matter at issue has already been tacitly conceded. Anyone who troubles to ask for a reason why nature should have a certain feature is thereby proceeding within a framework of thought where nature's rationality—the amenability of its features to rational explanation—is already presumed.

Ultimately the validation of optimalism lies in the very nature of the principle itself. It is self-substantiating, seeing it is automatically for the best that the best alternative should exist rather than an inferior rival for whose realization reason can devise no equally powerful case.[56] But this is just one of its assets; it also offers significant systemic advantages. For of the various plausible existential principles, it transpires—in the end—that it is optimalism that offers the best available alternative.

The principle being, as it were, self-explanatory, and for this very reason asking to ask for a different sort of explanation would be inappropriate. We must expect that any ultimate principle should explain itself and cannot, in the very nature of things, admit of an external explanation in terms of something altogether different. And the impetus to realization inherent in authentic value lies in the very nature of value itself.

Yet what is to be the status of a Law of Optimality to the effect that "whatever possibility is for the best is ipso facto the possibility that is actualized." It is certainly not a logico-conceptually *necessary* truth; from the angle of theoretical logic it has to be seen as a contingent fact—albeit one not about nature as such, but rather one about the manifold of *real* possibility that underlies it. Insofar as

---

[56] As noted above, other principles can also be self-substantiating, albeit not in a way that meets present purpose.

necessary at all it obtains as a matter of ontological rather than logico-conceptual necessity, while the realm of possibility as a whole is presumably constituted by considerations of logico-metaphysical necessity alone. But the division of this realm into real vs. merely speculative possibilities can hinge on contingent considerations: there can be logically contingent laws of possibility even as there are logically contingent laws of nature (i.e., of reality). "But if it is contingent, then surely it must itself rest on some further explanation." Granted. It itself presumably has an explanation, seeing that one can and should maintain the Leibnizian Principle of Sufficient Reason to the effect that for every contingent fact there is a reason why it is so rather than otherwise.

The Law of Optimality thus has a *raison d'être* alright. But one that lies in its own nature. For it is, in the final analysis, for the best that the Law of Optimality should obtain—that the best of reasons speak on its behalf. After all, there is no decisive reason why that explanation has to be "deeper and different"—that is, no decisive reason why the prospect of *self-explanation* has to be excluded at this fundamental level. After all, we cannot go on putting the explanatory elephant on the back of the tortoise on the back of the alligator ad infinitum: as Aristotle already saw, the explanatory regress has to stop somewhere at the "final" theory —one that is literally "self-explanatory." And what better candidate could there be than the Law of Optimality itself with the result that the divisions between real and merely theoretical possibilities is as it is (i.e., value based) because that itself is for the best?

We must expect that any ultimate principle should explain itself and cannot, in the very nature of things, admit of an external explanation in terms of something altogether different. And the impetus to realization inherent in authentic value lies in the very nature of value itself. A rational person would not favor the inferior alternative; and there is no reason to think that a rational reality would do so either.

After all, to achieve an adequate resolution of our ultimate question the principle at work cannot rest on further extraneous considerations. For the question of why the truth of things is what it actually is will arise with respect to the principle itself, and if it is to resolve such matters it must do so with respect to itself as well. It must, in short, be self-sustaining and self-grounding. Otherwise the requisite ultimacy will thus be achieved.

The question "Why optimalism?" splits into two decidedly distinct parts, namely (1) "Why is it that optimalism obtains?" and (2) "Why is it that *we should accept* optimalism's obtaining?" These issues are, of course, every bit as distinct as "Why did Booth assassinate Lincoln?" and "Why should we accept that Booth assassinated Lincoln?" The former question seeks an existence for a fact, the latter asks for the evidentiation of a judgment.

As already noted, the answer to the first question is straightforward. Optimalism obtains because it is self-potentiating. It is the case that what is for the best obtains because this itself is for the best. Optimalism, in sum, obtains on its own self-sufficient footing. Why should what is for the best exist? The answer lies in the very nature of the principle itself. It is self-substantiating, seeing it is automatically for the best that the best alternative should exist rather than an inferior rival. Value is, or can be, an explanatory terminus: it can be regress stopping and "final" by way of self-explanation in a way that causality or purposiveness can never manage to be.

Optimalism and Evaluative Metaphysics

Optimalism has it that even as truly rational people will perform *what they think is best* (not only for themselves but ideally for everyone concerned), so a rationally functioning reality will act so as to realize what is actually for the best, everything considered that for whose realization the best reasons obtain. There can be no good reason for reality to function otherwise—a circumstance that, after all, is inherent in the very idea of what a "good reason" is.

Optimalism has many theoretical advantages. For example, it is conceivable, so one might contend, that the *existence* of the world—that is to say, of *a* world—is a necessary fact while nevertheless its *nature* (i.e., of *which* world) is contingent. And this would mean that separate and potentially different answers would have to be provided for the questions "Why is there anything at all?" and "Why is the character of existence as is—why is it that this particular world exists?" However, an axiogenetic approach enjoys the advantage of rational economy in that it proceeds uniformly here. It provides a single uniform rationale for both answers—namely that "this is for the best." It accordingly also enjoys the significant merit of providing for the rational economy of explanatory principles. For to ask for a different sort of explanation would be inappropriate. We must expect that any ultimate principle must explain itself and cannot, in the very nature of things, admit of an external explanation in terms of something altogether different. The impetus to realization inherent in authentic value lies in the very nature of value itself. To reemphasize: a rational person would not favor the inferior alternative and there is no good reason why a rational reality should do so either.

But whence does the good obtain its creative impetus? Simply as part of the world's pervasive lawfulness. What is at issue, however, is not a law not (as yet) of nature but rather a proto-law of naturizing. Whence does mass get its power to convert to energy, to deflect space so as to engender gravity. That's just how things work. (And why should they work that way?—Just because that's for the best.)

There is nothing all that new and original in the idea that value (merit, being "for the best") exerts on existential impetus. There are innumerable traces of this line of thought in Plato and Aristotle, in Neo-Platonism, in Christian scholasticism, and indeed in some contemporary thinkers as well (e.g., John Leslie). What is new in the present discussion is only the web of argumentation by which this ancient idea of axogenesis is being supported.

From its earliest days, metaphysics has been understood also to include "axiology," the evaluative and normative assessment of the things that exist. Already with Aristotle the aim of the enterprise was not just to describe or characterize, but to *grade* (appraise, rank) matters in point of their inherent value. Such metaphysical evaluation has two cardinal features: (i) it is genuine evaluation that involves some authentic concept of greater or lesser value and (ii) the mode of value involved is *sui generis* and thus not ethical, aesthetic, utilitarian, etc. Accordingly, it evaluates types of *things or conditions of things existing in nature* (not acts or artifacts) with a view to their intrinsic merit (not simply their "value-*for*" man or anything else). The very possibility of this axiological enterprise accordingly rests on the acceptance of distinctly metaphysical values—as opposed to ethical (right/wrong) or aesthetic (beautiful/ugly) or practical (useful/unuseful) ones.

The paternity of evaluative metaphysics in philosophical practice can unhesitatingly be laid at Plato's door, but as a conscious and deliberate philosophical method it can be ascribed to Aristotle. In the *Physics* and the *De Anima* we find him at work not merely at classifying the kinds of things there are in the world, but in ranking and grading them in terms of relative evaluations. Above all, his preoccupation in the *Metaphysics* with the ranking schematism of prior/posterior —for which see especially chap. 11 of Bk. 5 (Delta), and chap. 8 of Bk. 9 (Theta)— is indicative of Aristotle's far-reaching concern with the evaluative dimension of metaphysical inquiry.[57] It was thus a sound insight into the thought-framework of the great Stagirite that led the anti-Aristotelian writers of the Renaissance, and later preeminently Descartes and Spinoza, to attack the Platonic/Aristotelian conception of the embodiment of value in natural and the modern logical positivist opponents of metaphysics to attach the stigma of illegitimacy to all evalua-

---

57 His willingness to subscribe to teleological/axiological explanation is clearly attested by Aristotle's account of the rationale of the continuity of organic existence: "For since some existing things are eternal ... while others are capable both of being and of not being, and since the good ... is always accordingly to its own nature a cause of the better in things ... for these reasons there is generation of animals." (*De Generatione Animalium*, 731b24–31).

tive disciplines. Nevertheless, despite such attacks, evaluative metaphysics has continued as an ongoing part of the Western philosophical tradition as continued by such thinkers as Leibniz, Kant, Hegel, and Whitehead, all of whom envision world-systems where some things have greater value than others.

A prime example of this methodological approach in recent philosophy is G. E. Moore's *Principia Ethica*.[58] For Moore taught that the realm of ethical values is not self-contained but rather roots in a manifold of metaphysical values. His celebrated "method of absolute isolation" invites us to make comparative evaluations of two hypothetical worlds supposed to be alike in all relevant respects except that in one of them some factor is exhibited which is lacking in the other. Thus Moore argues for the intrinsic value of natural beauty (i.e., its value even apart from human contemplation) by the argument:

> [A hypothetical] beautiful world would be better still, if there were human beings in it to contemplate and enjoy its beauty. But that admission makes nothing against my point. If it be once admitted that the beautiful world in itself is better than the ugly, then it follows, that however many beings may enjoy it, and however much better their enjoyment may be than it is itself, yet its mere existence adds something to the goodness of the whole: it is not only a means to our end, but also itself a part thereof. (*Op. cit.*, § 50)

To espouse the project of evaluative metaphysics is thus to give Moore the right as against Henry Sidgwick's thesis that: "If we consider carefully such permanent results as are commonly judged to be good, other than qualities of human beings, we find nothing that, on reflection, appears to possess this quality of goodness out of relation to human existence, or at least to some consciousness or feeling."[59] (There is of course the trivial fact if "we" do the considering, then "we" do the evaluating. But the point to be borne in mind is that this need not be done from a humanly parochial let alone an idiosyncratically personal and "subjective" standpoint.)

Sidgwick to the contrary notwithstanding, man is neither the measure nor necessarily even the measurer of all things in the evaluative domain. Things no more become valuable because we think them to be so than they are intelligible or efficacious because we deem them so. Along these lines, as Leibniz already saw, the value of an existential domain is determined by an optimal balance of rational desiderata such as procedural *order* (uniformity, symmetry) and phenomenal *variety* (richness, plenitude)—duly reflected in such cognitive features as intelligibility and interest. It is its (presumed) gearing to an inherently

---

[58] Moore 1903. See in particular §§ 50, 55, 57, 112–113.
[59] Sidgwick 1874, Bk. I, Chap. ix, § 4.

positive value which like economy or elegance, is plausibly identifiable as physically relevant and as such subject to scientific inquiry, that establishes optimalism as a reasonable proposition and ultimately prevents the thesis "optimalism obtains because that's for the best" from declining into vacuity. To be sure, this means that optimalism is not so much a *principle* as a *program*.

The Role of Intelligence

Optimalism obviously presupposes a manifold of suitable value parameters, invoking certain rationally-friendly and yet physically relevant features (symmetry, economy, uniformity, harmony, or the like) as merit-manifesting factors. And here optimization at issue is—and should be—geared to a "scientifically reputable" theory of some suitable kind, coordinate with a complex of physically relevant factors of a suitable kind. After all, many a possible world will maximize a strange "value" of *some* sort (confusion and nastiness included). It is its (presumed) gearing to a positive value which like economy symmetry, or stability is plausibly identifiable as physically relevant—contingently identifiable as such subject to scientific inquiry—that establishes optimalism as a reasonable proposition and ultimately prevents the thesis "optimalism obtains because that's for the best" from declining into vacuity. But this essential factor can and should—as we have seen—be provided in terms of the rational order that is maximally beneficial to the interests of intelligence.

In seeking to furnish an explanation of the why of the world's state of things we presuppose that natural reality meets the salient demands of explanatory intelligence. So if success in this venture can be achieved at all—as inquiring intelligence demands it must—then we might as well assume from the outset that reality is so constituted as to meet this demand, and thereby so constituted *as though* it were the product of an intelligent agent or agency. This being so, it would seem that the most plausible consideration for a holistic explanatory will be one to which any such theory must, if adequate, have to commit itself, viz. that the real is rational in exactly the manner that an intelligence-geared optimalism envisions.

Why should there be a link between rational existence and existence? Why should the merits affiliated to reason provide grounds for being, reality, and existence? Theology apart how could goodness possibly provide for reality? This is certainly a legitimate and appropriate question. It certainly cannot be omitted from the problem agenda. But when it is placed there it is already too late. For the question is itself predicated on a crucial presupposition, viz. that there will indeed be a rational existence explanation—a rationally appropriate account for a particular condition of things (why existence is as it is). But the very idea that there is such an account is predicated on the rationality of existence. If existence is to be subject to rational accounting then rationality must be an inherent feature of existence. (After all, if an A-process is to explain the existence of B then A-conformity must constitute a defining aspect of what B indeed is.) The very demand for the rational explanation of a fact itself betokens the rationality of the actual.

But there is also another way of looking at the matter. The schema

(1) A is the actual state of affairs because X

is predicated on the presupposition that there is such a thing as an existence-substantiating feature X. As argued above, this feature cannot be factual (then there would be circularity), and so it must be evaluative. To play its explanatory role X must differentiate A from any possible alternative: so it must be that A has more of X than any such A-alternative. So it must be a feature of positivity, because as the saying has it "Things could always be worse." So the scheme (1) comes to:

(2) A is the actual state of affairs because this provides for more of some suitably positive merit than any possible alternative.

But this is just exactly the Leibnizian "Principle of the Best."

This, to be sure, is an epistemic argument for *accepting* the principles, of explaining why we should adopt it. There yet remains the (very different) ontologically metaphysical question of validating *the principle*—of explaining why it is that what it claims is the case. But the answer to this question is provided by the principle itself. It is self-substantiating. The reason why it obtains is afforded by the principle itself: it obtains because this is for the best.

In sum, intelligence-geared optimalism affords a promising prospect for the explanatory resolution of holistic/synoptic questions because if any adequate and cogent grand explanatory theory is available at all, this cannot but mean that reality is so constituted as to merit the needs of inquiring intelligence.

In the end, then, there are basically two considerations that speak for the acceptance of optimalism: its ability to address ultimate questions positively in providing for a plausible explanatory transit form possibility to actuality; its ability to do this in a way that meets the actuality-abstractive demands of the situation, its capacity for self-substantiation. After all, in facing a choice among alternatives—be it in practical or in cognitive matters—we intelligent humans will, insofar as rational, opt for the best (most rationally cogent superior) of the alternatives. And an intelligence congenial reality is bound to do likewise.

Evidentiating Optimalism

So much for the ontological side of why optimalism obtains. But there of course yet remains the epistemological matter of its evidentiation—of why it is that one would be well advised to accept optimalism. This, of course, is something else again.

To obtain evidence for optimalism we will have to look at the world itself: the *substantiation* for optimalism will have to root in our knowledge of natural reality. And in *this* regard, optimalism, if tenable at all, will have to be tenable on at least roughly scientific grounds. And here the best evidence we could expect to have for optimalism is that emergence in the universe of intelligent beings able to understand the modus operandi of the universe itself: intelligent beings who can create thought-models of nature. For optimalism's evidentiation, then we must look to a universe that is user-friendly to cogency (intelligent) being in affording an environment that is congenial to the best interest of intelligence.

And so we confront the question "Is the world as we have it user-friendly for intelligence?" And the answer here would seem to be an emphatic affirmative with a view to the following considerations:
- the fact that the world's realities proceed and develop under the aegis of natural laws: that it is a manifold of lawful order whose doings exhibit a self-perpetuating stability of processual function;
- the fact of a course of cosmic development that has seen an ever-growing scope for manifolds of lawful order providing step by step the materials for the development front of the laws of physics, their theme of chemistry, their biology, their sociology, etc.;
- the fact that intelligent beings have in fact emerged—that nature's modus operandi has possibilized and facilitated the emergence of intelligence;
- the fact of an ever-deepening comprehension/penetration of nature's ways on the part of intelligent beings—their ongoing expansion and deepening of their underlying of the world's events and processes.

In sum, a substantial body of facts regarding the nature of the universe speaks on behalf of an intelligence-geared optimalism. Evidence is certainly not lacking here.

The Optimalistic Landscape

Optimalism is deeply rationalistic. It sees the optimal as being that for whose realization there is the best of reasons—that which would be actual if rationality had it is own way.

What sorts of considerations can possibly provide for the justificatory validation of optimalism? Why should it be that the Principle of Optimality obtains? Why should what is for the best be actual? But consider! To ask this question is to ask for a reason. It is *already* to presume or presuppose the rationality of things, taking the stance that what is so is and must be so for a reason. Once one poses the question "Why should it be that nature has the feature *F*?" it is already too late to raise the issue of nature's rationality. In advancing that question the matter at issue has already been tacitly conceded. Anyone who troubles to ask for a reason why nature should have a certain feature is thereby proceeding within a framework of thought where nature's rationality—the amenability of its features to rational explanation—is already presumed.

Ultimately the validation of optimalism lies in the very nature of the principle itself. It is self-substantiating, seeing it is automatically for the best that the best alternative should exist rather than an inferior rival for whose realization reason can devise no equally powerful case.[60] But this is just one of its assets; it also offers significant systemic advantages. For of the various plausible existential principles, it transpires—in the end—that it is optimalism that offers the best available alternative.

The principle being, as it were, self-explanatory, and for this very reason asking to ask for a different sort of explanation would be inappropriate. We must expect that any ultimate principle should explain itself and cannot, in the very nature of things, admit of an external explanation in terms of something altogether different. And the impetus to realization inherent in authentic value lies in the very nature of value itself.

Yet what is to be the status of a Law of Optimality to the effect that "whatever possibility is for the best is ipso facto the possibility that is actualized." It is certainly not a logico-conceptually *necessary* truth; from the angle of theoretical logic it has to be seen as a contingent fact—albeit one not about nature as such, but rather one about the manifold of *real* possibility that underlies it. Insofar as necessary at all it obtains as a matter of ontological rather than logico-conceptual necessity, while the realm of possibility as a whole is presumably constituted

---

[60] As noted above, other principles can also be self-substantiating, albeit not in a way that meets present purpose.

by considerations of logico-metaphysical necessity alone. But the division of this realm into real vs. merely speculative possibilities can hinge on contingent considerations: there can be logically contingent laws of possibility even as there are logically contingent laws of nature (i.e., of reality). "But if it is contingent, then surely it must itself rest on some further explanation." Granted. It itself presumably has an explanation, seeing that one can and should maintain the Leibnizian Principle of Sufficient Reason to the effect that for every contingent fact there is a reason why it is so rather than otherwise.

The Law of Optimality thus has a *raison d'être* alright. But one that lies in its own nature. For it is, in the final analysis, for the best that the Law of Optimality should obtain—that the best of reasons speak on its behalf. After all, there is no decisive reason why that explanation has to be "deeper and different"—that is, no decisive reason why the prospect of *self-explanation* has to be excluded at this fundamental level. After all, we cannot go on putting the explanatory elephant on the back of the tortoise on the back of the alligator ad infinitum: as Aristotle already saw, the explanatory regress has to stop somewhere at the "final" theory —one that is literally "self-explanatory." And what better candidate could there be than the Law of Optimality itself with the result that the divisions between real and merely theoretical possibilities is as it is (i.e., value based) because that itself is for the best?

We must expect that any ultimate principle should explain itself and cannot, in the very nature of things, admit of an external explanation in terms of something altogether different. And the impetus to realization inherent in authentic value lies in the very nature of value itself. A rational person would not favor the inferior alternative; and there is no reason to think that a rational reality would do so either.

After all, to achieve an adequate resolution of our ultimate question the principle at work cannot rest on further extraneous considerations. For the question of why the truth of things is what it actually is will arise with respect to the principle itself, and if it is to resolve such matters it must do so with respect to itself as well. It must, in short, be self-sustaining and self-grounding. Otherwise the requisite ultimacy will thus be achieved.

The question "Why optimalism?" splits into two decidedly distinct parts, namely: (1) "Why is it that optimalism obtains?" and (2) "Why is it that *we should accept* optimalism's obtaining?" These issues are, of course, every bit as distinct as "Why did Booth assassinate Lincoln?" and "Why should we accept that Booth assassinated Lincoln?" The former question seeks an existence for a fact, the latter asks for the evidentiation of a judgment.

As already noted, the answer to the first question is straightforward. Optimalism obtains because it is self-potentiating. It is the case that what is for the best obtains because this itself is for the best. Optimalism, in sum, obtains on its own self-sufficient footing. Why should what is for the best exist? The answer lies in the very nature of the principle itself. It is self-substantiating, seeing it is automatically for the best that the best alternative should exist rather than an inferior rival. Value is, or can be, an explanatory terminus: it can be regress-stopping and "final" by way of self-explanation in a way that causality or purposiveness can never manage to be.

In the end, the circumstance that the conditions of one problem constrains so radically nonstandard a mode of resolution as optimalism is a clear indication of the inherent difficulty of explanting totalization in philosophy. It poses a challenge that is both difficult and unavoidable.

11 Thematic Stability Amidst Philosophical Development

Preliminaries

Philosophizing can no more come to a conclusive standstill than can science. For both alike the unending variation of the evidential phenomena brought into view by the physical and conceptual technology of every age provides for an ever-changing panorama of individual and collective experience. And a rationality that calls for the alignment and harmonization of belief and experience is thus bound to yield an ever-changing perspective on philosophical issues. But nevertheless all of this change occurs amidst the stability of the fundamental problems and issues.

On casual inspection the history of philosophy thus makes the subject look to be a transit across an ever-changing landscape with old issues fading out of sight as ever new ones coming into view. But this impression is mistaken. In reality, the old issues morph into new shapes; the new issues pour fresh wine into old bottles. The development of philosophy is a story of change by attention and transformation rather than one of change by replacement and substitution.

That this is so can be illustrated by looking at the situation of Medieval Philosophy in its context of its theological orientation with the ontologically very different secular philosophy of the current era.

The Big Controversies of Medieval Philosophy

Medieval stochasticism has its own characteristic agenda of philosophical problems—often themselves largely derived from the Neo-Platonism of late classical antiquity. Primarily these include:

Universals

Is the classification of things into different descriptively kinds simply a matter of a human convenience in communication or does it represent differences in their natural constitution? Are the descriptive groupings of the world's constituents (types of things, spaces) the result of nature-imposed differences ("Platonism") or are they mere human contrivances devised for communicative convenience ("Nominalism")? Or do they represent an intermediary situation: the way in which material things strike us as we interact with them ("Conceptualism")? Is our descriptive language nature-corresponding, or purely a matter of human convention, or reflective of the specific modalities of our interactions with nature?

Future Contingency

Are the natural occurrences of the future pre-determined so that what will happen is already settled—our future human acts included. Or is there a contingency at work that precludes some actualities from being pre-necessitated. Are some future developments as yet undecided, or is contingency merely an artifact of our ignorance in that while it is determined what will happen, we just cannot know about it. (This comments to the theological issue of God's foreknowledge.) Are native occurrences pre-necessitated or are they indeterminately contingent, or is there an intermediate mix of some sort between these positions?

Free Will

Is there room for choice human affairs or is everything that happens in part of one all-determinative and coordinated plan? Are our human actions up to us or are they predetermined as part of God's plan? Do we authentically determine our ideas by free choice or is it all predetermined?

Divine Foreknowledge

Does God's omniscience means that realty spread out before God's mind I some vast and all-inclusive panorama. Does God know *everything* about the future? Or are the outcomes of free human choices outside his present cognition? Or is there an occasional indeterminacy where since *p* is as yet unsettled, failure to know it does not count as ignorance.

Averroism and Cognitive Authority: *Reason vs. Revelation*

Is knowledge one uniform and coherent whole, or are there two different realms of knowledge (say by natural inquiry and by super-natural revelation) which can potentially disagree with one another? Is truth one or dual and possibly conflicting?

Averroism roots in a recognition of the limitedness and finitude of human knowledge. Our access to reality is imperfect and we have to do the best we can within our limited recourses. Fortunality realizes these as a faith, transcendental resource, viz. revelation. Accordingly, human informativeness and the truths we accept fall into two categories: the one ordinary and based on observational information the other extra-ordinary and based on prophetic.

There are accordingly two realism of truths—that of everyday experience and its extremis is scientific inquiry, and are based on religious inspiration and revelations. By and large they can be brought into conformity. But in some matters across religion not only supplements but even corrects what we otherwise take ourselves to know. In matters of conflict the faithful accept religion, the doubtful philosophical judgement, and the skeptics adopt a minimalism confined to the deliverances of observational experience.

Ultimate Authority Regarding Duty and Obligation

Is the ultimate mandating authority in human affairs lodged in the Church (and so centered on the Pope) or in the secular power (and so centered on the king)? Or is the situation a hybrid with the king acting as God's agent or vice-regent on earth? And so there was a never-ending debate among royalists and churchman, regarding the crucial problem of what is owed to Caesar and what to God. Whose claims or finds a primacy, church or state. When the state taxes banquets, what about assets left to the church. When the state taxes imports (e.g. of wine) what about those destined for the church? Again there are three possibilities, depending on whether the functions of anthology in the church, the state, or a reciprocally accommodating mix.

Creation

Has the world always existed, with one version of it transferred from one previous one, or is there a *creation ex nihil* which provides an absolute starting point for physical existence.

Is the universe created (as per the Book of Genesis) or is it uncreated and eternal (as per the teachings of Aristotle). Or as we to conjoin the two as per an Averroist, dualizing approach.

Survival by Transformation

As theology-ignoring modernity came to prevail in the wake of the Enlightenment and "God is dead" thinking dominated philosophy an entirely different model of philosophizing came to the fore. A different set of issues now sets the stage and a different problem-agenda prevailed. Or so it might seem. But in actual fact, none of these controversies have vanished altogether from the philosophical scene. But they have not survived in exactly their medieval, theological form: all have undergone ongoing transformation as the State replaced the Church and then the Society the State as focus of philosophical impetus.

Thus the problem of universals is still with us, albeit not with abstract entities at its focus but rather meanings and their conceptual embodiment in of language. The ongoing issue is whether reality shapes and determines linguistic insight or whether this insight defines and determines what qualifies as such.

Again the problem of Free Will is still out there, unresolved. In essence it remains the same, albeit significantly reoriented. The literature of the subject is now greater than ever. But the issue is transformed—from foreseeability by God to predictability by science. But the issue now is whether it is the laws of nature rather than the will or understanding of God that precludes freedom.

Nor has the problem of the world's eternity left the philosophical scene. Originally it turned on the issue of the age of the universe: created or eternal? Now it turns on the question: did the Big Bang originate physical existence altogether, or did it merely inaugurate a new cycle in ongoing succession of events of cosmic history? Was the origin of this universe something absolute or merely new state or phase?

To be sure the, older conflict of God vs. Caesar and Pope vs. King no longer enjoys dominant primacy in an era where the pope is the CEO's of religious conglomerates and kings are figureheads. But the fundamental issue of law-stipulated vs. ethically mandated, normative appropriateness remains intact.

Again, the Averrosits doctrine of dual truth—biblical vs. philosophical, revelation vs. Aristotle—no longer presses upon us in its original format. But in a transformed version—humanistic vs. scientistic, spiritual vs biological it still survives as a matter of ongoing debate.

In effect all of those philosophical issues of the medieval problem-agenda persist after having undergone a "sea change." And this sort of situation has posted throughout the development of the subject.

Philosophical Development as Genealogical

The point here is that over time the inherent dialectic of philosophical development readjusts and realizing the fundamental issues of the changing conditions and circumstances of the time. Thus consider the following examples of such historical transformation:
1. Freedom of the will. What is the impediment to autonomous
   Antiquity: Fate
   Medieval: The will of God
   Modern: Natural law at the cosmic level
   Our Contemporaries: Brain physiology and psychological determinism at the persona level
2. The ultimate authority over human affairs. What is the lawgiver of social normativity?
   Antiquity: Fate
   Medieval: God's will and divine mandates
   Modern: The monarch
   Our Contemporaries: The society and social order
3. Why be moral?
   Antiquity: The rules of the state
   Medieval: God's commandments
   Modern: Society's demands for social order
   Our Contemporaries: The self: the interiority of humanhood
4. Ultimate decider of the right: What is the ultimate source and arbiter of legitimate demands upon people?
   Antiquity: The Will of the Gods
   Medieval: God via the Church
   Modern: The Conscience
   Our Contemporaries: The self-constituting psyche

As such illustrations indicate, the salient "big problems" of philosophy's agenda remain fundamentally the same but undergo a sea change as to detail to achieve accommodation to the intellectual ethos of the time. At bottom the salient issues thus remain. There is no issuing death certificates in philosophy.

This, then, is a compact overview of major problems in medieval philosophizing. And as the preceding account indicates, there is a clear polarity at work in all of these disputes. Each of them has a theological pole alignment to the teaching of the church, opposite a secular/humanist pole. And in virtually every case there is an untenable compromise position designed to provide some sort of reconciliation between opposing extremes.

Thematic Evolution

The reality of it is that in matters of philosophical fundamentals the problems of an earlier era are neither solved or dissolved. They do not fade into nothingness, vanish, and depart the scene, but rather persist in changed form. Thus not a single one of those problems that pre-occupied the Medieval theorists has disappeared from the agenda of philosophy—they all survive but in mutated form having morphed into different but still suggestably kindred favor.

As European thought moved from medieval scholasticism to the Renaissance, terms of separation formed between the traditionalists and the modernizers. The former simply accepted the traditional characterizations to themselves, and formed names for their opponents—heretics in religion (Erasmists in Spain) empiscasts in medicine, Marxists in public affairs.[61] These controversies can be traced across the cultural landscape at large and were particularly virulent in the evolution of universities.

Consider an example. In every age humans assert their identity as such views by separation from distinguishes between themselves and "the other": Romans from barbarians, our contemporaries from robots. The distinction between one's own familiar humankind and similar yet different other sorts of beings runs throughout the ages. But the detail of it is ever changing by transformed understanding about the sort of being one most fundamentally is. But yet the fundamental problem of self-definition and types differentiate remains the same. The issues are reconfigured and transformed, but yet presume a deep structural regularity that renders them different resolution to the same question: "What is there about me that makes me and my follow ("my type of being") into what it is that we most fundamentally are". The answer of different issues are very different. ("A community of persons living under one law," "children of the same God," "common members of a biological species"). But the fundamental question remains the same at its core.

Radical Discontinuity?

Against this thesis of the persistence of philosophical problems one might argue a particularly drastic counterexample: the pervasive expulsion of God from the Eden of philosophy.

---

[61] This can be verified in Rashdall 1845—and in more detail in Addy 1966.

Up to the era of Descartes, Spinoza, and Leibniz, God figured centrally in every philosophical position. But Immanuel Kant (1724–1804) wrought a sea change here: he was not a theist: he did not believe in God. Yet a large part of his classic *Critique of Pure Reason* was nevertheless devoted to the divinity. For as Kant saw it, while God did not exist as such, the *idea of God* was centrally important. For it provides crucial reference-point by contrasting knowledge as we humans do and can have it with a correct and proper understanding of reality. Without this contrast conception we could not correctly apprehend the nature of human knowledge and its limits. So while God does not exist as a being, the idea of God and his knowledge is a cognitively crucial contrast resource for understanding. Reality in itself—as only God could comprehend it—provides the ultimate article of fact and contrasts critically with *our* reality, as we humans do and must see it, given the consideration of the human mind. And so, in expelling God the Supreme Being from metaphysics, Kant resurrected God as the crucial point of reference in epistemology.

G. W. F. Hegel (1770–1831) wanted to do Kant one better: He wanted to eject from epistemology not only God but also Kant's fictional replacement for him, functional replacement, the inevitable constitution of the human mind. As Hegel saw it, these transcendental resources are not needed to sustain the objectivity of our knowledge simply because there is no such objectivity—no Reality independent of human contrivance to which correct thought must conform. All there is Appearance—opinions people have—but this is subject to formation by a culturally grounded development Spirit that drives the church of opinion through successive changes that replace the lesser by the ample, the inferior by the improved.

The ancient Greeks had divided philosophy into three parts:
- *Logic* (epistemology, linguistic analysis, rhetoric);
- *Metaphysics* (existence, reality);
- *Ethics* (morals, political and social philosophy).

Kant had exiled God from the second part; Hegel from the first. It remained for Friedrich Nietzsche (1844–1900) to do likewise with the third. His "God is dead" teaching in moral philosophy effectively completed the job of ejecting God from philosophy. Over time, the secularization of philosophy was complete. God was a thing of philosophy's past.

Or was he?

Consider the data of Display 11.1 presenting the comparison of Google hits in early 2021. Atheism fails to leave Theism in the shade, nor do the secular philosophers (Kant, Hegel, Nietzsche) prevail massively over the religious (Aquinas,

Newman, Barth). Comparison at the level of wider public concern and interest does not support a God-is-dead interpretation. The views of culture's avant gaurde need not reflect those of the public at large. In the larger contest for people's hearts and minds, secular philosophizing has not won anything like a decisive victory.

Table 11.1: An Interesting Comparison Sampling (late 2020) (Google hits in millions)

Secular		Religious	
Secularism	15	Theism	15
Kant	29	Thomas Aquinas	22
Hegel	30	J. H. Newman	55
Nietzsche	42	Karl Barth	23

So much for historical data, let us return to the conceptual situation.

Even when some of the heretofore central issues (be they cognitive or practical or ideological, or whatever) are set aside in the historical course of things, philosophy with its characteristic concern for "the reason why," is called upon to show that this is needful and appropriate. It has to address the matter—that same range of issues—from a variant point of view (viz. now con instead of the earlier pro). But this of course means that the same objects of concern are still on the agenda, with the same range of issues still in view. (Even those who reject free will or bourgeois morality have to say what this is and how it is supposed function.) When the physician changes the diagnosis, the same patient and the same symptoms remain.

The developmental continuity of philosophical thinking has important implications for the study of the subject's history. For it means that not only the *issues* with which earlier thinkers were concerned and the methods by which they deliberated about them, but even the doctrines they addressed continue to have present-day reliance. Unlike scientific subjects like physics or chemistry themselves, the study of theirs can be useful and informative with regard to issues on the agenda of the present day. The deliberations of the past can in significant measure shed light on the issues not only with regard to their origins but even regarding the substance of the agenda of the present.

12 Philosophical Disagreement and Orientational Pluralism

> No one philosopher's system can be acceptable to another without some modification. That each must reject the thoughts of others ... is due not to causes in taste and temperament but to the logical structure of philosophical thought. (R. G. Collingwood, *An Essay on Philosophical Method*)

The Scandal of Philosophy

Time and again over the centuries, philosophers have dwelt with dismay on the inability of their discipline to lay to rest the disagreements of the past and to reach fixed and settled conclusions. Philosophers have often cast envious sidelong glances at the sciences, with their demonstrated capacity to solve the problems and settle the controversies of the field providing for a continually increasing volume of findings about which a general consensus can be achieved.

Already in antiquity, the skeptics maintained the futility of philosophy on this basis:

> That nothing is self-evident is plain, they [the skeptics] say, from the controversy which exists amongst the natural philosophers regarding, I imagine, all things, both sensibles and intelligibles; which controversy admits of no settlement because we can neither employ a sensible nor an intelligible criterion, since every criterion we may adopt is controverted and therefore discredited.[62]

Indeed, the conflicting views of the philosophers were the very first of the five *tropoi* or skeptical arguments *(rationes dubitandi)* by which Sextus sought to support a skeptical position.

At the dawn of modern philosophy, Descartes complained as follows:

> I shall not say anything about philosophy, but seeing that ... it has been cultivated for many centuries by the best minds that have ever lived, and that nevertheless no single thing is to be found in it which is not subject of dispute, and in consequence which is not dubious, I had not enough presumption to hope to fare better there than other men had done. And also, considering how many conflicting opinions there may be regarding the self-

---

Note: This chapter was originally published in *The Review of Metaphysics*, vol. 32 (1978), pp. 217–251.

62 Sextus Empiricus 1933.

same matter, all supported by learned people, while there can never be more than one which is true, I esteemed as well-nigh false all that only went as far as being probable.[63]

David Hume deplored philosophy's chaotic lack of consensus in the following terms:

> [E]ven the rabble without doors may judge from the noise and clamour, which they hear, that all goes not well within. There is nothing which is not the subject of debate, and in which men of learning are not of contrary opinions. The most trivial question escapes not our controversy, and in the most momentous we are not able to give any certain decision. Disputes are multiplied, as if every thing was uncertain; and these disputes are managed with the greatest warmth, as if every thing was certain. Amidst all this bustle 'tis not reason, which carries the prize, but eloquence; and no man needs ever despair of gaining proselytes to the most extravagant hypothesis, who has art enough to represent it in any favourable colours... From hence in my opinion arises that common prejudice against metaphysical reasonings of all kinds, even amongst those, who profess themselves scholars...[64]

Moritz Schlick gave an incisive formulation to this same point:

> But it is just the ablest thinkers who most rarely have believed that the results of earlier philosophizing, including that of the classical models, remain unshakable. This is shown by the fact that basically every new system starts again from the beginning, that every thinker seeks his own foundation and does not wish to stand on the shoulders of his predecessors... This peculiar fate of philosophy has been so often described and bemoaned that it is indeed pointless to discuss it at all. Silent skepticism and resignation seem to be the only appropriate attitudes. Two thousand years of experience seem to teach that efforts to put an end to the chaos of systems and to change the fate of philosophy can no longer be taken seriously.[65]

The same litany of complaints echoes down the years time and again, complaints regarding unsettled issues, unending disputes, unconvincing controversy, and unachieved consensus. And so we come to the question which philosophers have faced time and again: Why is the achievement of these desiderata apparently beyond our reach in this domain? Why are the practitioners of the discipline chronically incapable of reaching a meeting of minds?

---

63 Descartes 1637, pt. 1.
64 Hume 1911, Introduction.
65 Schlick 1959. Compare also Kant 1781, Preface to the First Edition, A viii-x; and Peirce 1934, "The Fixation of Belief," vol. 5, sec. 5.383.

Explaining Discord

Interestingly enough, this very problem of philosophical disagreement itself exemplifies the phenomenon at issue. For in addressing this question, too, philosophers have achieved no settled consensus. Diverse alternative accounts have been proposed, primarily the following:
1. *Sociological explanations* such as the contention that the reward system of the discipline countervails against agreement. Substantial credit only accrues to one who launches out in a distinctive direction of his own; otherwise, a philosopher is merely an exegete or a syncretist, rather than "a real contributor to the subject." Philosophers do not agree with one another because no professional benefits can accrue from doing so.
2. *Methodological explanations* that see the historical failure to reach consensus as attributable to a fundamental deficiency in the previously used mechanisms of philosophical practice. Descartes saw the deficiency of earlier philosophers in their failure to achieve a proper *method* of investigation. Hume saw the vanity of prior philosophizing in its failure to uncover appropriate *first premises* as a basis of reasoning. Peirce and others saw the failure of consensus to result from the absence of agreed *definitions* of terms, resulting from the lack of an adequate theory of meaning. In general, the relevant line is that philosophers have failed to agree because they have heretofore made an incorrect approach to the bases of reasoning in their field.
3. *Eliminative explanations* that dismiss the entire discipline as illegitimate. The issues of philosophy are illegitimate and improper pseudo-problems. They are not real questions and do not admit of any sensible answer. Inquiry does not arrive at a stable consensus in this field because philosophical issues just do not admit of any significant resolution at all.

The present discussion will argue that none of these explanations gets the matter right. The ground of philosophical discord must be sought elsewhere: in the very *modus operandi* of philosophical inquiry. The nature of the enterprise is itself such that it is unreason able to expect consensus, and inappropriate to lament its absence. Its inability to achieve consensus is not a proper subject for regret and not a "scandal of philosophy" but a *fact of life*—an intrinsic and inevitable part of its very nature as an intellectual discipline. It is a feature of "the logical structure of philosophical thought" as Collingwood puts it in the motto passage.

When one takes the reasons for diversity and discord to lie in the very nature of philosophical inquiry itself, one might mean that they reside (1) in the *problems* or issues of philosophy, or (2) in the solutions or *theses* propounded with respect to them, or (3) in the *arguments* used to defend the propriety of these solutions. The present discussion will argue that the root cause of diversity lies in a *combination* of these three factors in that *philosophical issues are always such that arguments of substantial prima facie cogency can be built up for a cluster of mutually incompatible theses*. Philosophical argumentation is accordingly *nonpreemptive:* the existence of one cogent resolution of an issue does not block the prospect of an equally cogent basis for its alternatives; by positive argumentation an excellent case can be built up in substantiation of each of several mutually incompatible theses. It is the virtually characteristic feature of philosophy that its problems are such that eminently plausible arguments, arguments that strike the doctrinally uncommitted ear as having more or less equal cogency, can be built up on mutually incompatible sides of the issue.[66]

In philosophy, supportive argumentation is never alternative precluding. Thus the fact that a good case can be made out for giving one particular answer to a philosophical question is never to be considered as constituting a valid reason for denying that an equally good case can be produced for some other incompatible answers to this question. The diversity of philosophical doctrine is rooted in the pervasiveness of such *aporetic clusters,* as one may call them. Every philosophical problem thus admits of a variety of mutually conflicting solutions on whose behalf an impressively cogent case can be made out.

Rival Resolutions

Philosophical issues will accordingly center about a family of interrelated theses T_i, a family which is assertively *overdeterminative* in the sense of the conflict of inconsistency:

1. On the one hand, there is substantial reason to maintain each one of these collectively incompatible theses because each has "much to be said for it"— an impressive case can be built up for maintaining each one.
2. On the other hand, the bare demand of logical consistency requires the *elimination* of certain of these theses.

And so, philosophical issues always revolve about an *incompatible* n-ad of individually self-substantiable theses. This circumstance is illustrated by the following approximative example based on the traditional "problem of free will":

---

[66] For a discussion of relevant themes see Gallie 1964, esp. chap. 8 on essentially contested concepts, and Weitz 1978. See also Kekes 1979 which gives a vivid and insightful analysis of the issues, and compare "Gallie 1977, which offers an illuminating contrast between this author's views and those of Gallie.

T_1: A genuinely free act cannot be causally determined (for if there are individual causal determinants, then the act is *eo ipso* not free);
T_2: Men can and do make free acts of choice;
T_3: All human acts are causally determined.

It is clear that these three theses represent an inconsistent triad. A restoration of consistency requires the abandonment of one or another of these theses. Accordingly, consistency can be restored by any of three distinct approaches: the abandonment of T_1, T_2, or T_3, respectively. Three alternatives arise:
T_1-abandonment: "Compatibilism" (of free action ad causal determination);
T_2-abandonment: "determinism" (of the will by causal constraints);
T_3-abandonment: "voluntarism" (the exemption of free acts of will from causal determination).

This situation is paradigmatic in philosophical controversy. For example, a major family of controversies in moral philosophy revolves about the following inconsistent triad of theses:[67]
1. (Ethical Cognitivism) We have knowledge of certain ethical facts;
2. (Ethical Autonomy) Neither experience nor reasoning yields ethical knowledge;
3. (Empiricism) There is no source of knowledge apart from experience and reasoning.

Various major positions can be exfoliated from the alternative modes of resolution of this inconsistency. Thus *Ethical skepticism* reasons from (2) and (3) to the denial of (1). *Ethical intuitionism* reasons from (1) and (2) to the denial of (3). *Ethical naturalism* reasons from (1) and (3) to the denial of (2). Each pair of the cluster shares a common premise which the third alternative denies.

Or again, consider the traditional problem of skepticism regarding our knowledge of the external world, which can be articulated against the background of the following three individually appealing contentions.
1. One can only *know* what is altogether certain-that regarding which every possibility of deception is eliminated;
2. Certainty can never be achieved because one is never in a position to eliminate the possibility of deception by an all-powerful malicious being;
3. We know various and sundry facts about the world we live in.

---

[67] This example is taken from Chisholm 1977, pp. 123–24.

If one abandons (1), one opts for a *defeasibilist* conception of knowledge. If one abandons (2), one espouses a *common sensist* theory of certainty. If one abandons (3), one adopts *skepticism*. All three of these variations on this theme of conflict resolution are standard staple of epistemological fare.[68]

Its linkage to such an aporetic cluster is an invariable characteristic of an authentic philosophical problem.[69] Once a relatively straight forward mode of resolution comes to hand-once an unproblematically decisive case can be developed for one particular resolution-the problem emigrates from philosophy. The emergence of various special sciences by fission from philosophy illustrates this phenomenon.

This continual circulation within a family of interrelated issues lies at the core of the historical unity of philosophy: it establishes this unity as thematic and problem-oriented. *Philosophia perennis* is like science and unlike *belles lettres,* at any rate in this respect, that its unity is a matter of the linkage of topics, issues, and problems rather than one of tradition, approach, or other noncognitive styles of unity.

Philosophical issues are accordingly akin to the antinomies studied by logicians and mathematicians-the Liar Paradox (Epimenides) or Russell's Antinomy, for example.[70] These, too, have been scrutinized for many years by powerful minds with the result of discord and divergence. Everyone agrees *that* some sort of consistency-restoring resolution is necessary, but there is no consensus regarding *what* the appropriate resolution is because of disagreement as to which of the individually inevitable seeming components of the antinomy is to be sacrificed on the altar of consistency-restoration. Invariably, every proposed

---

68 It is helpful in this regard to consider in this connection the dictum known as "Ramsey's Maxim." With regard to disputes about fundamental questions that do not seem capable of a decisive settlement, Frank Plumpton Ramsey wrote:

In such cases it is a heuristic maxim that the truth lies not in one of the two disputed views but in some third possibility which has not yet been thought of, which we can only discover by rejecting something assumed as obvious by both the disputants. Ramsey 1931, pp. 115–16.

69 This linkage of a philosophical thesis to an aporetic cluster in which it stands linked in correlative apposition to its rivals makes it plausible to hold the paradoxical-sounding view that "the argument for [and against] a philosophical statement is always a part of its meaning." (Johnstone 1959, p. 32.) For the position at issue only comes to be defined as such in the context of the counterpositions it proposes to exclude; in philosophy Spinoza's dictum holds: *omnis determinatio est negatio.*

70 This analogy between the logical and mathematical antinomies and philosophical issues is the focus of a (successful) 1977 research proposal by Benson Mates to the J. S. Guggenheim Memorial Foundation for which the present author acted as referee, and on which he has drawn in some paragraphs of the present discussion. Of course, the fundamental thesis that the central problems of traditional metaphysics are antinomous or paralogistic goes back to Kant 1781.

"solution" to such an antinomy achieves the benefit of resolution at the cost of doing violence to our fundamental intuitions concerning the concepts at issue. The situation in philosophy is of essentially the same sort.

The structure of philosophical issues is thus such that positive argumentation in support of their resolution fosters rather than removes diversity. The initial phase of positive argumentation in substantiation of philosophical doctrines always results in conflicts of inconsistency. Someplace along the line something must be made to give way. A secondary, negative phase to eliminate some of the plausible alternatives established by positive argumentation thus becomes a crucial part of the venture. Eliminative argumentation becomes indispensable.

This line of consideration accounts for what is-on first view-a puzzling aspect of the field, namely, the prominence in the philosophical literature of counterargumentation and refutatory discussions. In mathematics no one troubles to argue that 14 or 32 are *not* satisfactory solutions to a certain problem. This would be pointless because the number of incorrect answers is endless. But when there are only a limited number of viable alternative candidates in the running, negative and eliminative argumentation will obviously come to play a much more substantial part.

There would seem to be one straightforward and natural way to proceed in the face of the inconsistent n-ad of propositions that comprises what we have called an aporetic cluster of theses that group themselves about a philosophical issue. This is simply to scrutinize the supporting arguments for each thesis and make a comparative assessment of their strength. This done, one simply proceeds to eliminate those with the weakest case. The governing rule is seemingly simple: *Break the chain at the weakest link.*

This tactic represents a straightforward-seeming approach-one that requires a simple *comparative* analysis of the relative strength of the various courses of positive argumentation. But this tactic runs into a decisive obstacle. In most fields of inquiry (mathematics, the sciences, etc.) the acceptability of a conclusion depends ultimately on the merits of the argumentation that produces it. But in philosophy the situation is different. Here the acceptability of the overall argumentation turns pivotally upon the conclusions to which it leads. The standard way of assessing a philosophical doctrine is not only through the acceptability of its basis (the principles that constitute the *terminus a quo* of its argumentation), but no less implicitly through the acceptability of its outcome (the consequences that afford its *terminus ad quem*).[71] For one cannot evaluate the

---

[71] As Bertrand Russell put it, where philosophical issues are concerned we have an inductive situation that yields "reasons rather for believing the premises because true conclusions follow

strength of a philosophical argument *independently* of assessing the overall plausibility—the acceptability—of the consequences which it yields. But this fact means that we *already* need to be in a position to assess the relative merits of the various theses that are a party to the controversy-i.e., to carry out the very task which this process is designed to accomplish.

Accordingly, a rather different approach must be developed to provide for the second-stage, negative or eliminative phase of philosophical reasoning. The structure of the situation is such as to call for such an essentially evaluative methodology for the assessment of the several (strictly intellectual) costs and benefits involved in the adoption or rejection of alternative doctrinal positions.

Cost and Benefits

The fact is that this eliminative phase takes the form of what is, to all intents and purposes, a form of intellectual cost-accounting, a cost-benefit analysis. Its *modus operandi* consists in the evaluation and appraisal of the various "costs" and "benefits" of the several alternatives, one which proceeds via a weighing of the standing of the alternatives vis-a.-vis various parameters of merit (and demerit). These include not only formal criteria like consistency, uniformity (treating like cases alike), comprehensiveness, systematic elegance, simplicity, economy ("Ockham's razor," etc.), but also various material criteria like closeness to common sense, explanatory adequacy, inherent plausibility, allocations of presumption and burden of proof, etc. Such standards of probative procedure serve to determine the methodological matrix of the discipline by affording a framework within which adequate solutions to its problems exist.

Methodological orientations (in the sense presently at issue) are not theories or doctrines. They do not involve the acceptance of theses or propositions regarding how matters stand, but rather embody judgments of plausibility, principles of presumption~ intellectual *predilections* as it were. These represent a certain *predisposition* to assign probative weight-to give credence, to accord cognitive preference, to value this rather than that structural feature of one's beliefs, etc.[72] In

---

from them, than for believing the consequences because they follow from the premises" (Preface to Russell 1910). Cf. Plato 375BC, 533.

72 There are, of course, going to be some methodological commitments that will be shared by several probative orientations and, indeed, some that will run across all of them. The inferential mechanisms of logic are presumably one example. Another is the disciplinary nature of reasoning which serves to determine just what it is to do *philosophy* as opposed to writing fiction for intellectuals.

sum, what is involved in such an orientation is not a matter of *theses* but rather one of *methods*—methods, to be sure, of a rather special sort, namely *probative* methods of inquiry and substantiation, and specifically the devices inherent in our methods of cognitive appraisal or evaluation. What is at issue are the different schedules of merit and demerit-different norms of plausibility, cogency, etc.—that we bring *ab initio* to the enterprise of argument appraisal. (Such norms of plausibility are not *theses* about what is the case, but the cognitive *values* at work in the probative process.)[73]

A probative orientation of the sort at issue here is *axiological*it embodies certain value predispositions regarding the probative appraisal of theses and arguments. It is a matter of the mechanisms and standards of argumentation-in particular, as regards plausibility and presumption, the allocation of benefit of doubt and similar instruments of probative procedure.[74] Accordingly, such a perspective is not an algorithm for producing solutions to philosophical problems given certain constraints, but rather serves to define the constraints themselves, forming a part of what is needed to mark a solution as acceptable, to see that the solution is a solution.

But whence do these orientational commitments regarding the *modus operandi* of philosophical argumentation come at the outset before philosophical reflection and systematization? From our culture, the "spirit of the times," the heritage of our teachers and their enemies, the course of our experiences, our own personality, even the stage of development of inclination of a single individual. They are not stable or intersubjective but variable and even to some extent person-relative.[75]

---

[73] The situation in representative art is somewhat similar. As Nelson Goodman has written in an analogous context:

Which is the more faithful portrait of a man—the one by Holbein or the one by Manet or the one by Sharukie or the one by Durer or the one by Cezanne or the one by Picasso? Each different way of painting represents a different way of seeing; each makes its selection; its emphasis; each uses its own vocabulary... Goodman 1972, p. 28.)

[74] For further detail regarding the relevant issues, see Rescher 1976, 1977, and 1978.

[75] William James characterized the differences at issue as ultimately temperamental in character. In his classic essay on "The Present Dilemma in Philosophy" he wrote:

The history of philosophy is to a great extent that of a certain clash of human temperaments. Undignified as such a treatment may seem to some of my colleagues, I shall have to take account of this clash and explain a good many of the divergencies of philosophers by it. Of what ever temperament a professional philosopher is, he tries, when philosophizing, to sink the fact of his temperament. Temperament is no conventionally recognized reason, so he urges impersonal reasons only for his conclusions. Yet his temperament really gives him a stron-

It is, however, important to recognize that such evaluative predispositions need by no means always *prevail* in our reasoning. One may be forced in a certain direction of credence in spite of and notwithstanding one's inclinations in another. For example, one may incline to trust some source, but bitter experience may ultimately undo this initial inclination. The presumptions at issue are like most legal presumptions in being defeasible-liable to be upset or reversed by sufficiently weighty counterindications. A probative orientation exerts a certain cognitive pressure, but its force is not infinite and irresistible but can be dampened and even ultimately deflected and redirected. Accordingly, such orientations are not necessarily something fixed and immutable. They can change as the result of changes on the objective side (changing experiences or changes in belief and cognitive commitment, for example) or of changes on the subjective side (changes in attitude or taste, for example).

The sort of orientational disagreement that underlies doctrinal disagreement on this approach is not itself *doctrinal*- that is, it is not a disagreement about material theses or contentions. It is regulative rather than constitutive; methodological rather than substantive. It relates to the procedural frame of reference, rather than to material that is emplaced within such a frame. It is a matter of the standards and criteria through which a solution to a question can be validated as acceptable-for example, whether an appeal to what is "natural" or what is

---

ger bias than any of his more strictly objective premises. It loads the evidence for him one way or the other... ("The Present Dilemma in Philosophy" in McDermott 1977, p. 363.)

This stance was echoed by James's pragmatist congener F. C. S. Schiller, who saw philosophical differences as rooted in differences of *personality* and pictured philosophical commitment as the "legitimate offspring of an idiosyncracy." (Schiller 1934, p. 10.) But such stress on temperament does not get the matter quite right. It overemphasizes nature at the expense of nurture, whereas our philosophical value orientations are rather acquired than innate (though differences in temperament doubtless play some role). Nevertheless, James's approach is a step in the right direction in its suggestion that the ultimate basis of disagreement is extratheoretical because it is *pre* theoretical. The rationale of the probative valuation at issue proceeds from a level of thinking that is "logically prior" to the reasoned adoption of a specifically doctrinal stance. Newman puts the point as follows:

[A]ll reasonings beginning from premises, and these premises arising (if it so happen) in their first elements from personal characteristics, in which they are in fact in essential and irremediable variance with one another, the ratiocinative talent can do no more than point out where the difference between them lies, how far it is immaterial, when it is worthwhile continuing an argument between them, and when not. (Newman 1870, chap. 8, pt. 3, sect. 10.)

The whole of the immediately following section of Newman's book is a magisterial analysis of how differences in approach to probative method make themselves felt in interpretative analysis.

"intuitively plausible" or what is "economical" or what is "simple" is to be allowed. The probative weight of various kinds of cognitive moves is at issue here. It is a matter of *values* (cognitive values, to be sure) rather than beliefs-of the proprieties of certain ways of handling the issues.

In sum, orientational disagreement is not so much a disagreement about facts as about values. *The orientations at issue are value orientations.* This is what makes the learning of philosophy not only a matter of mastering facts but also one of the acquisition of attitudes and "points of view"—a consideration that occasionally even endows the enterprise with a certain national and parochial coloration and serves to account for the gulf of mutual noncomprehension that separates various schools. (It is the comparative improminence of conflicting methodological orientations in science—i.e., the consensus regarding cognitive values in the domain-that decisively distinguishes science from philosophy and explains why the development of science is comparably discord-free.)

On such a view, then, disagreement in philosophy roots ultimately in divergent norms for evaluating the probative cogency of philosophical arguments.[76] Philosophical disagreement is basically axiological.

Tentativity

A philosophical position thus always has the implicit conditional farm: given a certain set of commitments regarding the relevant cognitive values, such and such a position and the question at issue is the appropriate one. It is a matter of resolving one way or another a choice among alternative methodological-evaluative commitments and probative issues. The question is one of cognitive axiology.

Our probative methodology is, after all, "logically prior" to our taking a doctrinal position. It serves as a determinant of which is cogent, and thus cannot

---

[76] In recent years, Peter Winch and his supporters have argued for a relativism of the standards of rationality, holding that beliefs and actions can only be regarded as rational within a framework of standards and criteria of rationality appropriate to the cultural setting in which those beliefs and actions arise. The *rationality* of cognition and praxis is always a framework internal issue in the sense of Rudolf Carnap's well-known distinction between internal and external questions. (See Winch 1958 and 1964; and for a survey of the resulting controversy see Esposito 1977.) The presently envisaged relativity of the value-standards of philosophical argumentation espouses at the level of the probative standards of different *philosophical* "schools" of approach a position roughly comparable to this sort of *cultural* relativity of the cogency-standards of rational deliberation.

(helpfully) be defended *ex nihilo,* without presupposing the standard of cogency it itself is to supply. Our probative value-orientation reflects the "paint of view" we bring to the controversy in terms of the instrumentalities we use to assess the "casts" and "benefits" involved in the retention and rejection of the theses and doctrines that constitute alternative philosophical positions in the choice forced upon us by the aporetic clusters at issue. We reach the position of what might be called an *orientational pluralism* in philosophy-a view which has it that there are different and *(in abstracto*—as concerns the general principles of the matter) equally eligible alternative evaluative orientations which underwrite different and mutually incompatible resolutions of philosophical Issues.

In such circumstances, it hardly makes sense to characterize the endeavor to resolve a philosophical question as "a search for the (uniquely) correct answer." If there are several incompatible answers, no one of which can in the abstract nature of the case be certified as categorically superior, the very conception of "the right answer" runs into difficulties. The best one can do in such a case is to establish one alternative as optimally tenable in the light of the standards of plausibility and appropriateness that constitute a certain preestablished probative-value orientation.

On such a view, it emerges that a philosophical thesis is unquantified "true" only relative to a particular probative orientation. Categorical truth would then have to amount to "true from *the true orientation"*—from the only demonstratively correct and appropriate one. But there is no. such thing as "the correct orientation" here. Orientations of the sort at issue are not theses at all; their status is rather methodological than substantive: they are alternative *approaches* to the processes of intellectual conflict-resolution—perspectives that we bring to problem situations.

The Goodmanian nominalist does not disprove the existence of sets as distinct from their elements-he *refuses to allow* that they can play any role in "adequate explanation." The Quinean realist does not refute the existence of possibilia—he *does not countenance* them in philosophical expositions. The explanatory recourse to such resources is viewed as a defect, a liability, a demerit. Their role is rather as a mechanism of appraisal than as an argued doctrine. With Goodman we have to "see" that abstracta are bad, with Quine that possibilia are-on the basis of considerations like simplicity, economy, elegance, etc. At bottom, the issue pivots on matters of evaluation rather than doctrine.[77]

---

77 A very different but yet structurally analogous approach to philosophical pluralism, namely the *axiomatic* approach, sees the differences in teachings among different schools-in their doctrinal positions-to rest ultimately in a difference regarding axioms: fundamental commitments, basic premises, first principles, or ultimate beliefs about the nature of the world. The root of dif-

But can one not deploy rational argumentation to constrain another to adopt one's own probative orientation by straightforward argumentation, by devising a more or less conclusive case for its adoption?

The answer is negative. The sort of methodological stance at issue in a probative-value orientation does not involve any theses/ claims on the same level of discussion. It does not *make* any assertions-though it does provide a basis on which we ultimately incline towards certain assertions rather than others. As we have seen, it is a matter not of doctrinal commitments but of evaluative attitudes, of probative methods rather than substantive theses-of one's "intellectual predilections," as it were. The divergencies at issue with different orientations are ultimately evaluative and one cannot dispute about values from a value-free standpoint. One can, of course, reason (i.e., philosophize) about these evaluative stances, but this process is itself a stance-presupposing issue.[78]

To be sure, we can take an evaluative stance towards the spectrum of alternative probative orientations themselves: that is, one can evaluate such evaluative methodological orientations as being more or less appropriate or adequate, more or less comprehensive or systematically pleasing. But in philosophy there is no theory-neutral starting point, no fulcrum for an Archimedean lever that is itself outside the arena. The appraisal of a philosophical argument can thus only be carried out "from within," so to speak, that is, from the vantage point of a prior probative position from which the plausibility of the theses and inferences at issue can be assessed.

An objection looms. If our commitment to a particular probative orientation is not doctrinal in character-if it does not consist in the espousal of a set of substantive theses that are true or false—then it is not something that is rationally

---

ferences is seen as residing not in methodology but in divergent fundamental thesis-commitments. (For an interesting development of this position, with its emphasis on ultimate commitments—in the very broad sense of this term—see Johnstone 1959. The present approach does not take this line. For we reject the conception of ultimate theses in the domain at issue. It would be altogether unphilosophical to characterize a commitment to philosophical theses in terms of ultimacy. One must reject the very idea of "ultimate commitments" in philosophy the espousal of theses one is not prepared to defend. As Hegel quite properly stressed, there can be no axioms in philosophy. In rejecting the axiomatic approach as inappropriate in philosophy, we also reject the particular account of pluralism that accompanies it—an account that is, in its general structure, much like our own.

78 To be sure one probative value-orientation can be "superior" to another in the scope of its relevancy considerations and admissibility conditions. (Cf., e.g., the narrower sympathies of logical positivism with the ampler scope of Hegelianism or Ernst Cassirer's cultural neo-Kantianism.) But the "loss" or "gain" at issue can only be seen as such from an existing value-orientation that is already in place.

defensible, but is merely adopted "unquestioningly," and this surely renders it indistinguishable from an article of faith. It would thus appear that all philosophical inquiry rests ultimately on an irrational commitment.

Two observations are in order with respect to this objection. (1) The evaluative commitment that is ultimately at issue here is, to be sure, *unreasoned*, but it is not *irrational*. Rather, it is prerational; it implements the schematic idea of rationality by setting up the norms and standards by which the mechanisms of rationality do their work. (2) In philosophy, as in other modes of inquiry, we cannot justify our methodological orientation a priori-in advance of the conduct of inquiry—but we can retrovalidate (i.e., retrospectively revalidate) it in terms of the orientationally indicated overall acceptability of the results to which it leads. The structure of validation is thus rather cyclic than linear.

Pluralism

This pluralistic aspect of the situation endows philosophical controversy with a peculiar structure. It means that philosophical argumentation cannot rationally *constrain* consent save intraprogrammatically among those who share the basic commitment to a particular "probative-value orientation." In philosophy, we cannot really lock horns regarding fundamentals if we do not agree on he methodological first principles that form the framework of argumentation.

All that one can do in philosophy is to view the issues from one or another of a limited number of "available" methodological orientations. And the crucial fact is that *the very nature of a probative orientation is such that one cannot occupy several of them at once*. To be sure, in philosophy one can "shift orientations"—but the transition is so rendered that when the new orientation is attained the old one is lost. The shift involves a sort of intellectual conversion experience-one cannot retain the old values alongside of the new ones. An *axiological* (rather than *logical*) incompatibility obtains.

Descartes' proposal to overcome philosophical conflict by a recourse to *method* envisaged the plausible sequentialism of settling on a method first, and then using it to resolve conflicts. It is not difficult to see that this strategy will not work. After all, the probative orientations at issue are themselves methodological, and so the question of the relative merits cannot be settled by treating methodology as a neutral arbiter. As Hegel saw long ago, this approach is ineffective because the question of method is itself a philosophical issue: the methodological mechanism by which our discussion is conducted is itself a party to diversity and dispute. Controversy and argumentation about method itself is a part of philosophical work. We must recognize that pluralism also operates at the methodological level: the situation here is the same as in substantive philosophy. Neither method nor any other alternative provides a "neutral" (uncontroversial, unproblematic) point of entry into philosophical controversy.

Philosophical inquiry is such that whereas each mind will indeed arrive at *its* solution (with sufficiently hard work), still, nevertheless, no one solution can rationally construe the adhesion *of* all rational minds, regardless of their "mental set"—their methodological stand point of probative-value orientation. In these circumstances we must admit and heed the difference between: (1) *a globally* correct (or true) solution that is demonstrably right in itself, by its very nature (and is thus correct from every point of view, so to speak), and (2) a *locally* optimal (or adequate) solution that is cogent for those committed to a certain probative-value orientation. And given that the second perspective is apposite, it follows that such "demonstrations" as there are in philosophy do not have an absolute but only a relative force. In effect they take the form: IF you are prepared to make certain procedural commitments (i.e., to adopt a certain particular probative value orientation), THEN you will arrive at a certain particular solution. A rational constraint is clearly at issue here, but it is basis-relative (or orientation-relative), rather than absolute. Yet it is not a matter of "anything goes"—an orientation of the sort at issue is a commitment to a certain probative standard, and this notion itself has normative dimensions: not *any* sort of random, hit or miss process of endorsement affords a "probative standard."

The existence of pluralism in philosophy is thus inescapable. With respect to philosophical issues there always exists a variety of ("incompatible") solutions. We have no choice but to "agree to disagree."

Diversity and pluralism in philosophy can be accounted for in terms of a variety of explanatory models:

(1) *The Complex Reality View.* In philosophical systematizing we are dealing with a vast and complex reality which we (intellectually feeble) creatures simply cannot manage to grasp whole. We overemphasize what falls within the scope of our limited experience.

Diverse systems describe reality variably because they describe it partially (incompletely) in different aspects or regards. *Everybody is right-but* only for his own respective limited domain. An overarching complex reality embraces them all. Reconciliation between diverse doctrines can be effected *conjunctively* through the formula "but furthermore in *this* regard"—even as the elephant is spear-like in respect of his tusks and rope-like in respect of his tail. William James's pluralism is of this general sort:

> There is nothing improbable in the supposition than an analysis of the world may yield a number of formulae, all consistent with the facts. In physical science different formulae

may explain the phenomena equally well-the one-fluid and the two-fluid theories of electricity, for example. Why may it not be so with the world? Why may there not be different points of view for surveying it, within each of which all data harmonize, and which the observer may therefore either choose between, or simply cumulate one upon another? A Beethoven string quartet is truly, as some one has said, a scraping of horses' tails on cats' bowels, and may be exhaustively described in such terms; but the application of this description in no way precludes the simultaneous applicability of an entirely different description.[79]

Clearly when there are different aspects which we propose to combine-to "cumulate one upon another"—then we have in view the aspectual and mere-part-of-a-whole conception of the relationship between our systematizing and its object. Traditionally, this perspective is much favored among philosophers, who (beginning with Aristotle) like to think of the divergent views of their predecessors as so many partial insights into their own overreaching position fragmentary views which can all be coopted and reconciled within a wider universe of discourse (presumably that provided by themselves).

(2) *The No Reality View.* In philosophical systematizing we are dealing with an illusion. Theorists do not agree with one another because there's nothing to agree about-they are chasing a chimera. *Everybody is wrong.* The whole enterprise is based on the erroneous presupposition that the so-called "problems" of philosophy are sufficiently meaningful to admit of such a thing as "a correct solution." The positivist dismissal of metaphysics or a Kantian insistence on limiting ourselves to the confines of experience alone, holding that in transgressing these limits we enter the sphere of vitiating paralogisms, represent positions along roughly this line.

(3) *The Unique Reality View.* In philosophical systematizing we deal with unisolutional problems-problems where one solution is correct and all the alternatives are wrong. Only one party to a dispute is right-everybody else is just barking up the wrong tree. It's all a matter of having the right method (or perspective or axioms or whatever)—which presumably we ourselves do and others don't. Many of the classical metaphysicians inclined to this position. As William James put it:

> If we look at the history of opinions, we see that the empiricist tendency has largely prevailed in science, while in philosophy the absolutist tendency has had everything its own way. The characteristic sort of happiness, indeed, which philosophies yield has mainly consisted in the conviction felt by each successive school or system that by it bottom certitude had been attained. "Other philosophies are collections of opinions, mostly false; *my*

---

[79] McDermott 1977, p. 325.

philosophy gives standing-ground forever,"—who does not recognize in this the key-note of every system worthy of the name?[80]

(4) *The Perspectival Reality View.* One and only one position is right *from a given perspective of consideration.* But there is a plurality of perspectives. Various discordant perspectives do indeed exist, but none of them is "the right one," none can lay claim to being uniquely correct. Reality is multifaceted-regarded from different perspectives of consideration, philosophical issues demand different and discordant resolutions. And so, nobody is right *simpliciter* and unqualifiedly, because "right *simpliciter*" is an *inapplicable* conception here. Everybody (who is sufficiently careful and workmanlike) is right from the perspective of his own methodological approach.

Our present orientational relativism is akin to the last of these positions. It roots in the view that in philosophy we are dealing with real issues that admit of real (and unique) solutions-albeit solutions that are only attainable through approaching the issues from the vantage point of a commitment to a definite probative orientation (evaluative methodology).

When actually doing philosophy our standpoint is that of model number three, but when doing *metaphilosophy*—so that our own commitments are put into suspension (or *bracketed* in Husserl's terminology)-then the approach is altered. The position of model number four is now the natural one. But of course at this point-at the point where our own evaluative position is bracketed and where we adopt a deliberate "suspension of disbelief"—we are no longer doing philosophy as such.

But is number four really all that different from number one? Is the difference here not simply a matter of approach rather than of position? Are these not simply ways of describing what are-in the final analysis—precisely "the same facts"?

No. As long as we are *serious* about philosophical inquiry we must actually have a perspective of consideration (take an evaluative position, assume a methodological orientation). And here "internally" from our own point of view, so to speak-we cannot consider others as genuinely on a par with our own. Once a probative orientation is adopted-and adopted it must be-only one "correct" answer is ever available.

In philosophy we cannot say that "the real truth" is what holds from *every* methodological perspective. Nothing does. We cannot say that "the real truth" is what holds the *canonical* perspective (the correct one, the one at issue in

---

[80] James 1897, sec. 5.

the philosophers' penchant for the myth of the God's eye view). For only God knows what this is: there is no way for us to come to it. (Even if He were to tell us, the question of whether it is indeed He who is actually doing the telling would at once become a moot point for philosophical controversy.) And we cannot say that the real truth is what holds from *some* perspective i.e., at least one of the diversified spectrum of available possibilities. For it is rationally incongruous to opt concurrently for incompatible alternatives. The best we can do on behalf of our own solutions to philosophical issues is to claim that they afford "the truth *as we see it*," yielding a position that is bound to be accepted as correct by those who share our basic commitment to a particular probative-value orientation.

Variant Orientation

If orientational pluralism affords the proper diagnosis of the disease of philosophical controversy, then what is its cure-what is the appropriate stance to take on its basis? Several alternatives offer themselves.

The Skeptical Route

> For every argument/doctrine there is some equally good counterargument/doctrine that balances it out. The entire enterprise is accordingly *futile*. Since there is no hope of getting at "the real truth" our best course is to suspend judgment. We would do well to abandon the whole project and go on to something more profitable.[81]

Thus Rudolf Carnap, for example, has argued in this way that the unattainability of an absolute framework for the evaluation of philosophical controversies entails the futility of the whole enterprise:

> It seemed important to me to show that many philosophical controversies actually concern the question whether a particular language form should be used... I wish to show that ev-

---

[81] The ancient skeptics regularly argued that we must suspend judgment where different perspectives yield conflicting results:

Since, then, all apparent objects are viewed in a certain place, and from a certain distance, or in a certain position, and each of these conditions produces a great divergency in the sense-impressions, as we mentioned above, we shall be compelled... to end up in suspension of judgment. For in fact anyone who purposes to give the preference to any of these impressions will be attempting the impossible. (Sextus Empiricus 1933, bk. 1, sec. 121.)

eryone is free to choose the rules of his language and thereby his logic in any way he wishes. This I called "the principle of tolerance"... In this way, assertions that a particular language is the correct language... are eliminated, and traditional ontological problems... are entirely abolished.[82]

But a critical shortcoming of this position lies in its failure to admit the seriousness of the issues. Can we really in the final analysis rest *content-rationally* content-with such an abandonment of this domain? Our intellectual stake in the questions at issue is simply too great to allow us to be comfortable in abandoning this whole inquiry.

The Positivist Route: Dissolutionism

Philosophical problems are paralogisms/antinomies. The very fact that they lack the prospect of any categorical and standpoint-free resolution means that they are *ipso facto* improper and illegitimate. They must be dismissed holus bolus as mere pseudo-problems and the whole enterprise of philosophy, as traditionally concerned, is inappropriate and irrational.

The defect of this position lies on the surface. To take the absence of a unique, demonstrably correct solution as basis for dismissing the whole enterprise bears the aura of petulance. We are unwilling to play the game because we cannot have it all our own narrowly conceived way-a stance reminiscent of Aesop's fable of the fox and the grapes.

The Indifferentist Route

In philosophy we find that there are-as it were-many plants in the garden, elaborating rather diverse life forms. There are thus many different exhibits in the intellectual museum. You pays your money and takes your choice. At bottom, all the alternatives are of equal value, so it doesn't really matter which one we pick. We simply choose this or that, making an essentially indifferent selection (indifferent at any rate from the rational point of view: there may be historical or sociological or psychological constraints).

This position too seems misguided. Relativistic pluralism does not or need not-underwrite indifferentism. If philosophical problems generally admit of several distinct solutions-each of which is optimal from a certain methodological

---

82 Carnap 1963.

orientation (rather than univocally correct)-this does *not* mean that it is a matter of idle indifference which one accepts; that one can simply choose at random or toss out because "it makes no difference." There is no globally (universally) correct diet, but that does not mean it is a matter of indifference what a person eats. There is no globally (universally) correct language, but that doesn't mean it is a matter of indifference what grunts and noises a person uses in his endeavors to communicate with others. It is not a matter of "anything goes"; internal diversification does not entail indifferentism.

The Orientational Route

This position-our own-is based on recognizing the questions as *meaningful,* the issues as *important,* and the inquiry as *legitimate.* The project is a serious one, and it is important that in pursuing it we do the very best we can. But this does not mean that we can attain monolithic solutions in orientation-independent ways. Yet while accepting pluralism and relativism, we nevertheless stress their limits by noting the restricted variety of viable alternative probative-value orientations.

As orientational relativism sees it, it is crucial that with respect to philosophical issues only a *limited* plurality of tenable positions will be available. One accepts that conflicting approaches each have much to be said for them, but insists that this fact that a philosophical issue has a limited number of distinct (and incompatible) solutions does not mean that there are no rational constraints at all. A pluralism of perspectival orientation is not a chaos: It simply means that there is not one unique demonstrably correct solution of universally cogent stringency. It is perfectly in order to say that given a cognitive-value perspective one and only one resolution is (contextually and constitutionally) correct.

A Dual Stance

Orientational pluralism takes a dualistic, two-tier stance.
1. It envisages one-sidedly dogmatic commitment "from where one sits" *at the basic doctrinal level-that* from one's own perspective (the only one that one has) there is one optimally adequate position.
2. It envisages a pluralistically relativistic recognition *at the metadoctrinal level* that other perspectives "are available" (but only to others!).

This two-tier approach requires us to do our philosophical work at two levels: (i) the basic level of substantive philosophical ideas, and (ii) the higher level of metaphilosophical deliberations.

At the substantive level of philosophical inquiry we can work out our own favored position, and develop "our own stand" on the issues, even while realizing-when we step back from this level to the metalevel to survey our own work (and its rivals) with due detachment that other positions also emerge as tenable for "other points of view."

At the higher level of metaphilosophical consideration we can, however, take a broader perspective that encompasses also other alternatives alongside our own. At the metatheoretical level we react to pluralism not with an indifferentist choice of A-or-B, but with an alternative-canvassing survey: A, B. Such a stance recognizes that "our truth"—the truth as we see it-is part of a larger complex that includes conflicting positions. William James was quite right in stressing the inherent pluralism of philosophical endeavor:

> *The* Truth: what a perfect idol of the rationalistic mind! I read in an old letter-from a gifted friend who died too young-these words: "In everything, in science, art, morals, and religion, there *must* be one system that is right and *every* other wrong." How characteristic of the enthusiasm of a certain stage of youth! At twenty-one we rise to such a challenge and expect to find the system. It never occurs to most of us even later that the question 'what is *the* Truth?' is no real question (being irrelative to all conditions) and that the whole notion of *the* truth is an abstraction from the fact of truths in the plural...[83]

But all this willingness to allow alternatives only happens "at a remove" when we step back from engagement in substantive philosophy and assume a metaphilosophical stance where our own perspectival commitments are suspended and "bracketed." (This is the approach of the detached *historian* of philosophy and of the deliberately impartial *expository teacher.*) At the substantive level our own "cognitive stance" is determinative-and determinative of what we unhesitatingly maintain as the truth. (We assert it seriously and all assertion is assertion *as true.*) Despite our recognition that they are "limited," we are perfectly serious about our own ground-level contentions (exactly as with the situation of the so-called "Preface Paradox," arising when an author apologizes in the preface for the mistakes he is sure are committed in the text itself). Life being what it is, the existence of contrary views need never prevent the rational man from articulating with confidence and conviction those philosophical views which represent how matters stand "to the best of our knowledge and belief."

---

[83] McDermott 1977, p. 450.

It is necessary to deny categorically and with vehemence that the orientational pluralist in metaphilosophy cannot be serious about working out his own position within the field. For even though philosophical issues do not admit of a uniquely adequate solution *an sich,* there still remains the fact of an *optimal solution for us,* that is, for someone who shares our "perspective"— our sense of the plausible, etc. We can and should realize that even if we do not prevail in general and *contra mundum* (as we indeed cannot do), we can still work out a good solution to such problems *relative to a certain basis* of orientational agreement. Even the orientational pluralist can say that he who abandons "dogmatism" in philosophy is a traitor to the spirit of the enterprise, because this is what good philosophical work demands at the substantive level. In abandoning universalistic claims for his position, he can perfectly well say:

> I take philosophical inquiry seriously because I am eager to discover what qualifies-from where I stand-as the only cogent and correct answers to these questions. That these should also prove to be *your* answers would be very nice but is nowise necessary. N or would its failure daunt my own confidence. After all, I do not feel the less persuaded of the rightness of my own standards of judgment in moral or political matters because I realize they are not shared by everybody.

This line of consideration leads to the central thesis of the orientational relativism. For it is the very *raison d' etre* of the theory to maintain *that-and* to show *how-*with respect to philosophy *the combination of a base-level monism with a metaphilosophical pluralism is an available and indeed an attractive position.* Orientational pluralism enables us to have it both ways, so to speak. For on its teaching we can and should work out our own answers to philosophical problems (in a way that is rationally sound and altogether cogent relative to our own methodological perspective of consideration), but nevertheless we need not thereby feel compelled to dismiss as mistaken and misguided the work of colleagues whose conscientious labors lead them to other solutions. We can be fervent in our attachment to our own position without writing off as altogether worthless the work of our competitors in the field.

Orientational pluralism thus becomes an available and a *useful* position at the Metaphilosophical level-a position nowise incompatible with a convinced monism at the base-level of philosophical inquiry itself. Philosophical partisanship is not incompatible with metaphilosophical tolerance.

Self and Others

It is not only possible but appropriate at once to work out what is the appropriate answer to philosophical problems-as best we can determine it "from where we sit"—*and nevertheless* to recognize that other viable solutions are available and need to be reckoned with. Accordingly one should regard in a rather different light the question of the task of the individual inquirer and the mission of the enterprise at large in its historical and communal aspect. The individual philosopher must do the best he can to elaborate and substantiate his solution to a philosophical issue. Nevertheless when he "steps back," so to speak, and detaches himself from his own methodological commitments by "bracketing" them, he can and should recognize *at the metaphilosophical level* that his own position is merely one alternative among others (a very privileged alternative to be sure, in being-as he sees it-the *correct* one). The work of the individual is *monistic,* that of the community *pluralistic.* We must needs recognize (and perhaps rejoice in) the fact that we cannot expect that our own favored views will finally prevail, and that "the communal mind" of philosophical inquiry will and must press beyond our own positions, that the market is too big to be cornered.

Putting the matter in this way makes it sound as though we should be prepared to welcome-rather than lament-pluralism and diversity in philosophy. And this is indeed so. For there are powerful *methodological* reasons why we should welcome the fact that the arena of philosophy presents a scene of discord rather than consensus.

In philosophy, one cannot perform actual experimental tests as one can in the natural sciences; we cannot put our theories to the test by experiment and observation. Science develops as a dialectic of interchange between theorists and experimentalists, philosophy as a dialectic of interchange between theorist and countertheorist. The only way we have of testing our philosophical doctrines is to expose them to the trial of counter-argumentation in an endeavor to see how, cogently objections can be met. Can putative difficulties be overcome naturally and economically (given our basic commitments), or does the defense of the position call for its constant amendment through qualifications that heap epicycle on epicycle to the point of a complexity that threatens the plausibility of the whole project? In philosophy, probing criticism is a form of compliment-an indication that one is willing to take the discussions at issue sufficiently seriously to utilize them as a testing-ground for forming one's own views. Criticism and controversy are thus a resource of value-a prime source of stimulus towards the development and refinement and substantiation of our position. The way to "progress in philosophy" lies through opposition-through a Toynbee-reminiscent dialectic of challenge and response. Opposition becomes a form of collaboration; it is to be welcomed rather than regretted, seeing that in philosophy headway is in large measure made by injecting new sophistication into old controversies so as to raise the level of discussion.

In the long run, to be sure, any given contribution to this dialectical work will eventually become obsolete in the light of subsequent criticism and refinement. Ultimately its value (however large) will become "merely historical" in that it could no longer be looked upon as a contribution to the *current* "state of the art" in the discussion of the issues. (The ongoing dialectic of objections and replies, of refinements in the light of new approaches, means that the work of philosophy can never be completed, that it is not feasible that anyone should ever be in a position to "have the last word" on a philosophical issue.)

It is crucial to keep in view the difference between "my philosophy" and "the enterprise of philosophy at large." The nature of philosophical work renders it necessary and proper that the stance of the individual be monistic-that he himself "take a position" for articulation, elaboration, and advocacy. But the stance of the community is pluralistic-it is not feasible, and not desirable, that any one particular position should prevail across the board. From the angle of the philosophical community as a whole, the shape of the discipline is bound to be that of a mosaic of discordant positions. It is not necessary, nor even possible, that a consensus be reached. Proper accomplishment in philosophy viewed as *a field* of inquiry calls for a division of labor that requires each era to have its Platonists and Aristotelians, its nominalists and conceptualists and universalists, its empiricists and its rationalists, its skeptics and its dogmatists, etc. At the metatheoretical level we must regard the philosophical enterprise in terms of a prismatic multisidedness that provides each worker with collaborators and adversaries and endows the communal process of philosophical work with an interactionism of creative tensions that makes for the dialectical movement of an advance across the whole doctrinal spectrum.

This pluralistic paradigm places the overall discipline of philosophy-regarded in a synoptic metaphilosophical perspective into a characteristic light. Philosophy as a communal venture becomes a matter of the competitive, and yet quasi-cooperative, endeavor to build up as good a case as possible for a diversified spectrum of discordant possibilities. What counts as crucial from this overreaching metaphilosophical standpoint is not a matter of "getting at the absolute truth," but rather of enhancing the quality of the argumentation and gaining a deepened understanding of the structure of alternative positions.

There is a crucial shift here from definitive *truth* to situational *adequacy*—at any rate from the metaphilosophical point of view. Not *correct* solutions but more (or less) adequate and *workmanlike* ones are the objects of our striving.

(Craftsmanship is a key component of the proper framework of appraisal in philosophy.) One can perfectly well admit as well wrought the case for a position we cannot accept. No abandonment, no indifferentism is at issue here, but a perennial struggle that yields not final solutions but broader and deeper insights.

The operation of diverse probative orientations means that the failure to achieve a settled state of philosophical opinion is not only one of disagreement as between one historical era and another, but also spells the failure to realize consensus within each successive era. The rival "schools" are (or could be) at work at every stage of the game. The evolution of philosophy is not a matter of linear development, with each successive stage a unified synthesis of all that has gone before. It is a matter of the concurrent development of discordant traditions and points of view in ongoing rivalry—a perpetual struggle between divergent approaches developing in continual opposition and apposition, destined never to peter out in the dull harmony of general agreement. The "noise and clamor" of philosophical disagreement is a formative component of its fate.

In philosophy we just aren't going to get consensus, and the search for it is a Quixotic quest. The goal of ultimate consensus is a legitimate regulative ideal for scientific inquiry, a domain where diverse probative orientations in methodology do not, or need not, play a prominent role. But the situation is quite otherwise in philosophy. The proper mission of the overall philosophical enterprise is not the attainment of a general consensus, but the development of a debate of high quality.[84]

One acute philosopher has offered the following suggestion:

> [W]hile building a new systematic theory [in philosophy], the author is already aware of its incompleteness and the revisability of its very basis ... [This] opens room for a plurality of coexisting systems and a dialogue among them. Relativism is avoided, however, because all those more or less subjective philosophical constructions ultimately acquire their meaning and value in a confrontation which is objective and invariant for all philosophies-the confrontation of *real* human beings and the *real* world which ultimately, a posteriori, shows best what are the practical implications of the one philosophical orientation or the other.[85]

---

[84] Someone might object:

We cannot afford to abandon the quest for consensus. For man must implement his theoretical beliefs in action, and at the level of action and praxis the agreement of men working in common for agreed ends is crucial to the furtherance of human interest.

This standpoint badly distorts and/or oversimplifies the matter. In practical (even political) matters we can disagree in beliefs and yet find areas of consensus and compromise. We can often "sink our differences" and agree on actions to be taken to the mutual advantage in the common interest. It is a crucial-and very fortunate-fact of social life that *agreement is not an indispensable requisite* for fruitful collaboration in matters of praxis.

[85] Markovic 1976, see p. 283.

This author then proposes the familiar appeal to pragmatic considerations to enforce monism in a pluralistic situation where theoretical considerations do not suffice for this end. There are good reasons to think that this strategy will not work: (i) Many or most philosophical debates relate to issues that are relatively remote from practical affairs. Bishop Berkeley would be the first to deny that an immaterialist is committed to principles of bridge building that would serve to differentiate him from an atomist. (ii) Even if pragmatic considerations *did* militate for one doctrine over against another, the adherents of the latter will presumably disdain to give them weight, holding that the object of philosophical inquiry is to grasp the truth, not to feather our nests. (iii) More fundamentally, the very idea of pragmatic efficacy (benign practical implications) as a desideratum itself is an issue of controversy-it is a party to the dispute rather than a neutral arbiter capable of settling it.[86]

Pragmatism accordingly does not qualify as a destroyer of relativistic pluralism in philosophy.[87]

Operations and Problems

Let us consider briefly some lines of objection to the pluralism of the present theory.

1. *Does relativistic pluralism entail skepticism?* It might seem that a metaphilosophical pluralism must inevitably lead to skepticism regarding philosophical issues. But the matter is not quite so simple. The pluralism at issue maintains that there is indeed such a thing as "the real truth" with regard to philosophical issues, but that this "real truth"—the only truth there is in this domain—is orientational and inherently variegated. The stance of the theory is not that philosophical truth is unattainable let alone nonexistent, but that it is something inherently many-faceted. And there is no question of *indifference* here, once we assume the posture of a particular probative-value orientation (as indeed we must). Pluralism does not entail skepticism.

---

[86] This situation holds a fortiori when the standard of appraisal in terms of problem-solving capacity is shifted from practical to theoretical contexts. For here it is not just the *solution* to the problem that is theory-relative but even the very problem itself-Le., the question of the extent to which a certain issue poses a genuine and substantial problem.

[87] In the domain of *scientific* knowledge, however, the situation may well be-and indeed in the author's opinion is-crucially different in this regard. See Rescher 1977.

2. *Is orientational relativism defeatist?* Orientational pluralism holds that there is no *categorically correct* answer to a philosophical question, but only a (limited and circumscribed) spectrum of *perspectivally appropriate* answers.

But is it not defeatist to give up on the search for "the definitive truth"—the "uniquely correct" answer in a transperspectival sense? Surely not! If the situation is indeed as we have portrayed it, pluralism is not defeatist, but simply realistic-just a matter of recognizing "the facts of life."

3. *Does orientational pluralism support irrationalism?* Does not the step of abandoning the idea of "the ultimately and uniquely true and correct answer" in exchange for a variety of "perspectivally appropriate answers" mean an abandonment of the search for absolute truth and thus spell a lapse into irrationalism?

Not at all! For one thing, it is eminently rational rather than the reverse to refrain from asking for what cannot in the nature of things be had. On the view being advanced here it follows that relativistic pluralism is inescapable given the logical structure of the situation—a situation where the unavailability of a *unique* solution is simply an inevitable fact. The course of rationality is to accept the enterprise as rational inquiry shows it to be.[88]

Moreover, it must be recognized that our theory maintains various more straightforwardly rationalistic commitments: It holds that from the theory-internal standpoint of a given methodological orientation, it is (or should be) subject to undebatably cogent demonstration that there is only one single unique best resolution of a philosophical issue. And it holds that the ("external") facts *about* the orientations at issue are matters of decisive demonstration; in particular, the hypothetical fact that IF one espouses a particular orientation, then certain results follow. Moreover, the number of workable alternative solutions for which a good case can be made out (for *any* rationally available perspective of consideration) is finite: there is always only a limited spectrum of rationally viable positions.[89] Finally, and most importantly, the probative commitments and

---

[88] It warrants note that orientational pluralism endows the enterprise of metaphilosophy with a large and (hopefully) fertile research program centered around the exploration of ways in which one methodological orientation will be interpretable by approximative methods within others. (It is, as we have seen, crucial that what a philosopher is doing can be agreed to be of the genus "justification" and "problem-resolution," even by those who reject the specific probative value-commitments of his particular orientation.)

[89] This consideration both explains and is confirmed by the peculiar nature of "progress" in philosophy. For such progress is generally made by going back to some essentially identical earlier position and reworking it—articulating it with greater sophistication and defending it with greater ingenuity against counterarguments (which themselves become increasingly refined in the course of such dialectical interchange).

evaluative constraints that make up a perspective must be *intelligible*. And this intelligibility itself must be perfectly objective and orientation indifferent, in that the *operation* of the commitments and constraints at issue must be understandable to the partisans of other orientations[90] and not just understandable in themselves, but understandable (i.e., interpretable) as adding up to a viable methodological orientation. Thus all philosophers, regardless of their commitments in point of probative methodology, can still argue with shared disciplinary constraints about whether what somebody says or writes can count as "doing philosophy" or is just talk (or talk of some extra philosophical sort). In sum, our pluralistic theory is replete with rational limits and constraints.[91]

4. *Does orientational pluralism entail that philosophy is pointless?* For is it not pointless to strive to resolve a problem if one cannot arrive at a categorically correct (absolutely true, intersubjectively valid) answer, but only a relatively appropriate answer from the standpoint of a certain probative-value orientation? If *that* is how the matter stands-on such a footing of orientational pluralism-then why not abandon the whole enterprise?

The best tactic is to take the bull by the horns and to meet this question with a counter-question: Why should one be dissatisfied with a *relatively* appropriate answer to a philosophical issue, one that can be shown by cogent argumentation to be appropriate "from where one stands"—on the basis of the probative orientation one does in fact hold? If we can work out a resolution that is demonstrably cogent given our own actual perspective of consideration, then why not rest content? Surely there is a perfectly good point to the enterprise of finding what is (given my orientation-and that of those who share it) a demonstrably appropriate solution. It makes good sense to set about the work of finding answers which afford us with the satisfactions of rational cogency, even if they may well leave somebody else dissatisfied. Why should one succumb to delusions of grandeur

---

90 This is a much weaker claim than that of translatability. The sort of normative/evaluative *modus operandi* of probative procedure at issue in a "methodological orientation" is rather different from the fundamentally *descriptive* mechanisms at issue in a "conceptual scheme." The idea of alternative conceptual perspectives has been called into question in Davidson 1973–74. However, Davidson's negative argumentation turns too narrowly on considerations of *translation,* to the exclusion of paraphrase, circumlocution, and other merely *approximative* rather than *reproductive* devices for translinguistic transfer. But be this as it may, the idea of alternative conceptual schemes and that of alternative probative devices are distinct and separable.

91 Some of the substantive methodological discussions presented in other writings by the author give a more detailed picture of the mechanisms of plausible reasoning, cost-benefit analysis of cognitive problems, etc., that are deployed by a philosophical perspective in determining optimal resolutions to philosophical issues. See Rescher 1976, 1977, and 1978.

in philosophy and insist that everyone adopt one's own positions? As long as its potential inappropriateness from *someone else's probative orientation* nowise invalidates or undermines the appropriateness and cogency of a given position *from mine,* why should this mere lack of universality occasion discontent?

5. *Must isothenia produce epochê?* The ancient skeptics took the stance that a situation of "balancing out" obtains in most cognitive disciplines-philosophy included. For every argument in favor of some dogmatic contention some intrinsically no less cogent, they held, equally plausible case can be built up on the contrary side. Every pro-consideration can be balanced off against a con-consideration. This balance-or equipollence-of *pro* and *con* they designated as *isothenia* ("equilibrium").

The skeptics took the stance that the only rationally appropriate response to such situations of balancing-out is a suspension of judgment (*epochê*). When the pro-considerations are cancelled out by con-considerations, and conversely, total abstinence from any endorsement or rejection is-so they maintained-the only defensible course. In circumstances where equally cogent counterarguments can be advanced against any position, a suspension of judgment is the rational policy.

This line of thought is clearly correct when mutually annihilating pro- and con-considerations lie on the same level of discussion. But this is *not* the case in the context of the present deliberations.

For our position has it that at the ground-floor level of philosophical controversy-where we do and indeed must possess a "probative value orientation"—a situation of indifference or balance never obtains, seeing that such a perspective has it that one solution is emphatically superior to another. There is no balancing-out *isothenia* here, at the basic level of the inquiry itself.

To be sure, when we "step backwards" (so to speak) from the arena of controversy and abstract from our commitment to a particular probative orientation-where we move from the basic level to the metalevel-then we must acknowledge that a comparably plausible seeming case can be built up for other, conflicting positions. But this sort of second-level *isothenia* which, as it were, *brackets out* methodological commitments, certainly does not constitute a cogent reason for a suspension of judgment *(epoche)* at the basic-floor level of philosophical deliberation itself. It cannot be emphasized too strongly that a metaphilosophical recognition that there are other positions which are *in abstracto* available-albeit only from the angle of an orientation that one certainly does not share oneself-constitutes no reason whatsoever to give up a firm and dedicated adherence to the philosophical position to which one stands committed.

In one way, however, the position of the skeptic does indeed apply. In view of the existing situation of the field, it does seem quite inappropriate to claim philosophical *knowledge*. We might indeed claim to know that a certain philosophical argument is fallacious, or perhaps even that it is valid; but it sounds incongruous to claim that one knows the conclusion to which it leads. In science one might say things like "Whereas people used to think... it is now known that... is the case." But we do not talk like that in philosophy. One might indeed say that "Nobody nowadays seriously espouses Plato's theory of ideas" or "At this time of day, there are no longer adherents to Aristotle's theory of final causes." But no careful writer would say "Plato's theory of ideas has been shown incorrect" or "We now know that the theory of final causation is wrong." Whereas in science one can quite properly say "Galen's theory of the origin of disease through humor-imbalance is nowadays known to be just plain wrong," a conscientious philosopher would not say this sort of thing about Gassendi's occasionalism or Berkeley's immaterialism.

In philosophy certain positions and theses may be rejected as ill-argued, as unpalatable, even as bizarre-sounding to the contemporary mind. But the discipline is such that it is incongruous-sounding and a sign of a braggadocio that immediately puts any knowledgeable person off-to say things like "It is now known that...," "It has recently been established that...," "X has shown that...," and the like. Philosophical claims can quite properly be made in the language of plausibility, and no doubt even in that of warranted assertability. But it is a token of the green novice or the pompous fool to advance them in the language of knowledge and demonstration in the language of established certainty, rather than the more modest language of reasoned opinion. The skeptic is right to this extent that the operation of *isothenia* at the metaphilosophical level means that here we have a field where actual knowledge must not be claimed in addressing a community, large sectors of which will generally fail to share one's own methodological orientation. But to say that the absence of *epistêmê* cancels out altogether the rational credentials of the discipline would-as we have seen-be a gross misreading of the situation.

We should no more expect uniformity in philosophy than on anything else in the sphere of thought and endeavor. Disagreement and difference is the practice or differences in culture and context. Chinese philosophers can be expected to differ from the Greeks even as their contentions or their diets differ. But such contextuality is not a matter of indifferent relativism, it is rationally enjoined by differences in fundamentals.

In a rational endeavor conclusions require premises, claims require substantiation. These materials for reasoning are ultimately applied by experience in the broadest sense of that term. Our philosophical position must ultimately be guided in the manifold of experience, And experience is a framework of context—temporal, cultural, social—and even to some extent personal From the same premises in the same contextual nature, rationality will enjoin the same conclusions, But it no less requires that with different inputs different outputs will be obtained. Doctrinal differences not in irrationality or fecklessness but in the variation of experiential context.

Is Orientational Pluralism Self-Consistent?

One critical nettle must be grasped. Is orientational pluralism not self-refuting? For does it not imply that other discordant, contrary views regarding the roots of philosophical controversy are *also* available? Yes-of course it does. And of course they are. A reprise of the pluralism inherent in orientational relativism is encountered at the metaphilosophical level, where a spectrum of familiar alternatives lies before us in the diverse methodologies that underlie project abandoning skepticism, positivist dissolutionism, relativistic indifferentism, and so on. And all of these have much to be said on their behalf. The orientational relativist unhesitatingly faces up to this fact. He grants that there are other metaphilosophical approaches which are also somehow or other "available." But he insists that from where he sits (from *his* methodological stance) this fact is actually confirmatory. Given the orientationalist approach, such pluralism at the methodological level is exactly what one would expect to encounter, since metaphilosophical considerations are themselves philosophical in character. An orientational pluralist is gladly willing (and perfectly able) to view the theoretical availability of rationally cogent approaches alternative to his own-and the seeming plausibility of some of them-as a point of *substantiation* of his theory rather than as a *refutation* thereof. For in philosophy as elsewhere we can only move forward from where we presently are.[92]

---

[92] This view is reminiscent of Protagoras's reported contention that every issue can be disputed with equal validity on either side, including even this issue itself: Protagoras *ait de omni re in utramque partem disputari posse ex aequo et de hac ipsa, an omni res in utramque partem disputabilis sit.* (Seneca 1863, 88, 43.)

13 Philosophical Cogency

Philosophizing

Philosophy as an endeavor in answering questions and solving problems. As such it is a quest for knowledge in the guidance of action—a normative inquiry into what should be thought and done by us. From the outset its deliberations address three first-order issues:
1. *Knowledge* (Fact): What should we accept [by way of contention as to how things stand in this world and how they function there]? Here we are concerned with rational inquiry in the workings of reality: in short with the unified production of science. *(Epistemology)*
2. *Evaluation* (Worth, Merit, Value): What should we prize and cherish? Here we are concerned with rational inquiry into the evaluation of the real and the possible. *(Axiology)*
3. *Action* (Choice and its Implementation): What should we do: how should we comport our actions and efforts in this world?; and how avail ourselves of the opportunities that existence has put at our disposal? Here we are concerned with rational inquiry the problems of managing our affairs. *(Ethics)*

Throughout, philosophy is a rational inquiry into how human endeavor to achieve understanding of our condition as intelligent agents.

And finally there is yet another level of deliberations to be considered over and above the prevailing *what, why,* and *how* deliberations. This is the realm of consideration which encompasses and unifies everything has gone before—looking to our knowledge of effective action for the realization of value—namely *metaphysics*, the heart and core of the philosophical enterprise, the domain of deliberations where every point of the descriptive come together in a collaborative attempt to address the central problem in which the very reason for being of the discipline is encapsulated.

Finally, there is the higher order, second-order of consideration, namely that of *metaphilosophy*, which addresses the rationale and the methodology of philosophy itself.

To answer philosophical questions one needs information, and this only becomes accessible to us as the product of investigation and deliberation. And in philosophy as in science inquiry requires data, although very different sorts of data are at issue. For in science the data are hard—the fruits *observation and measurement*. In philosophy they are soft, the fruits of *speculation* and even of what is merely plausible supposition.

Accordingly, philosophical reasoning proceeds in a manner very different from and in a way opposite to the mode of reasoning in natural science. Science begins with the hard facts of natural or artificial (experimental) observation and proceeds to devise explanatory theories to account for them. In philosophy we begin with soft data and seek to reconcile and harmonize them. Science thus moves ampliatively from facts to explanatory plausibilities. Philosophy, by contrast, moves reductively from seeming plausibilities to their most harmoniously coordinated subsector. Science strive to extend and amplify its data, while philosophy diminishes and constricts them: science is inductive philosophy constructive and reductive.

The Data of Philosophy

The totality of philosophical data as described in the preceding chapter is too much of a muchness: taken altogether they conflict and are all too obviously subject to discord, disagreement, and inconsistency. Like the deliverances of sight or hearing these data too can malfunction and go awry. But they enjoy a presumptive status that can only be set aside for good reasons. And it is to them that we must look for evidentiation of our more speculative forays into less commonplace issues. Philosophy must therefore deploy a systematic screening—a winnowing that separates the wheat from the chaff (so to speak). Cognitive prioritization with regard to acceptability is its characteristic procedure. Its preceding is not one of instructive application but one of inductive narrowing that makes the less plausible give way to what is more so.

When there is conflict and discourse in information some items must give way to others. And this reductivity is definitely characteristic the methodology of philosophizing, a circumstance that is only seldom apparently because philosophers almost always present only results and maintain evidence regarding the procedure by which they arrive at them.

Consider an example. Political economists have told us such things as:
1. Property is theft: The privacy of goods violate the rightful claims of the wider community whose operation have those goods available (Marxist dictum);
2. Theft is wrongdoing it is morally and legally inacceptable to take from others what they own (ethical precept);
3. Moral wrongdoing should be abolished: people should be deflected from not be morally inappropriate practice (social doctrine);
4. Private property is appropriate and ought to be socially accepted and its diffusion enhanced (common belief)

But it is clear that these four data-constituting theses are logically incompatible: (1)–(3) taken together entail not-(4). In the interest of logical coherence one of them at least has to be abandoned.

And so four alternatives lie open:
- *Rejection 1:* One could maintain that property acquisition can arise by productivity in ways that do not violate the due claims of others—that when I write a poem, bake a cake, plant a crop, or the like, others have no claims on these products;
- *Rejection 2:* One could maintain that moral convention gets it wrong—that when the thief need for something is greater than its current owner's he is justified in taking it (think of Victor Hugo's *Les Miserables*);
- *Rejection 3:* One could maintain that moral wrongdoing cannot be rejected flat-out because its practice can often benefit the material interests of the community (think of ROM *Fable of the Bees*);
- *Rejection 4:* One could maintain—as per Marxism—that private property above and beyond personal needs (clothing, eyeglasses, medications, etc.) should be abolished in the interests of social benefit.

Each of these rejections provide for a particular philosophical position. And in the mere logic of things rationality demands that endorsing at least one or the other of these fours rejections becomes rationally unavoidable in the face of that incompatibility. But in doing this we must, as rational inquirers, have some ground or basis to justify taking that seeing this particular item as the place to break the chain of inconsistency. And so as we try to make our way back to cosmology we confront different possibilities or procedures, each definitive of a different philosophical approach. The interplay of contradiction and rationality propels us into philosophizing.

How is one to assess the plusses and minuses (the pros and cons) on a proposed issue-resolving answer to a question? Basically by means of for meta-questions:
1. Is the answer well grounded? Is there social evidence for its tenability?
2. Does the answer itself have problematic presuppositions?
3. Does the answer pose further questions that are difficult to resolve? Does it open up problematic issues and raise more (or larger) problems than it solves?
4. What are the larger implications of the answer? To what extent does it help with the resolution of other relevant problems?

In sum, two considerations are paramount in assessing the merit of a proposed problem-resolution, namely its evidentiation and substantiation, and its problem-field reduction by resolving (or helping to resolve) more difficulties than it itself introduces.

Granted, this issue-resolution itself puts a big item on the problem agenda: viz. how to assess the magnitude of problems. In this regard it is clear that one must distinguish between the *real* magnitude (whatever that is!) and their *presently apparent* magnitude. And it is clear that all that we can possibly manage to work with is our judgement regarding the second of these.

Philosophical Criticism

The idea that criticism is a central issue in philosophy originates in Immanuel Kant's classical *Critique of Pure Reason*. And subsequent developments afford many variations of this theme. In particular there are the topics of:
– Criticism of philosophy;
– Criticism in philosophy;
– Criticism of a (particular) philosophy;
– Critical philosophy (as a distinctive project).

Let us consider each in turn.

Criticism of philosophy. This usually takes the turn of a so-called "positivism" which holds that philosophy is an obsolete and inappropriate enterprise whose self-proclaimed aim of understanding reality has been replaced by the positive sciences. This position effectively self-destructs because its scientistic attempt to show that everything knowable can be established by science is itself not a scientifically validatable thesis.

Criticism in philosophy is a matter of employing a critical methodology, which, in seeking to detect problems and difficulties in working out a philosophical position, looks to ways of overcoming or bypassing them. It implements a constructive method of inquiry, which though a critical search for weaknesses in a position, looks to ways of diminishing their impact. A position's strength lies not in the circumstance that it there require no critical analysis, but rather in the fact that it is able to meet objections in one way or another. The crux here is not immunity from objections but the capacity to meet and survive them.

Criticism of a (particular) philosophy is an essentially adversarial procedure. It is a matter of identifying weaknesses of so serious a nature that the tenability of a position at issue is compromised. Usually this way of proceeding is adopted with the aim of defeating a rival practice in the interests of supporting one that is favored. Such criticism generally seeks to establish that a philosophical position is subject to a particularly damaging sort of flaw, primarily including inconsistency, oversimplification, conceptual unclarity, productive insufficiency, myopia (in ignoring crucial issues), and others. Unlike constructive criticism in philosophy, this mode of destructive criticism seeks out weakness in order to defeat the position at issue.

Critical philosophy. Here criticism is a mode of investigation for establishing the range and limits of a particular approach. It examines a doctrine critically in order to determine the limits of its tenability and by way of seeing how far it can be extended without encountering insurmountable difficulties. No philosophical position is immune from such criticism because the issue of its scope and limits invariably needs to be addressed. If it has limits, these must be demarcated; if it has none, this must be shown.

Such critical philosophizing prioritizes the Kantian vision of critical assessment as an instrumentality for accessing the scope and limits of cognitive resources.

For criticism in philosophy as sketched above, the basic idea is that of *constructive* criticism—spotting weakness and identifying grounds of objection in the interest of meeting and overcoming them. And unlike "literary criticism" whose aim is (or should be) to illuminate the aesthetic qualities of its object of concern, this philosophical criticism is aimed to establish its tenability.

On this basis, philosophical criticism is not a matter of making finding fault for negativism's sake, but becomes a constructive instrument instrumentality of the lack of establishing the merit of a position's credentials and claims. It is thus a crucial instrumentality of the rational investigation of philosophical problems and thereby a mode of collaboration in the development of a presentation.

Philosophical Substantiation

As philosophers seek to consolidate their favored approach to considering maintenance, what entitles them to expect that their readers will accept their contentions? In the final analysis the answer will have to pivot on answering the question: "How serviceable is it from my own point of view? To what extent does it square with the realities of my experience?"

Granted, someone might not have an expectation of acceptance, and simply not care. The attitude may be "These are my views, take them or leave them." Then of course the issue falls by the wayside: where there is no intent to convince—to convince by reasons—then the life of reason is abandoned. And now it becomes questionable whether the discussion, however interesting, qualifies as philosophy. For when you simply tell me your thoughts without further informative substantives we are dealing purely with biography not inquiry. And so when the question of rational acceptance—of cogent grounding—is to the agenda, the issue of substantiation and evidentiating grounds becomes rationally unavoidable. It is inherent in the rational scheme of things that you can expect one to accept what you do only insofar as I have grounds for doing so. At this point the thoughtful exploitation of those data becomes critical. Here, however, the philosopher has little choice but to appeal to his interlocutors course of personal experience. The question here: What manifold of commitments best conforms to and agrees with the course of your experience? At this point philosophizing becomes impersonal in method.

Contextualism vs. Relativism

The philosophical empiricists were hung up on sense experience. Forget that! Human experience is broader and richer than the sensory. But philosophizing is indeed conditional and limited by the range of experience. And the horizons of our knowledge are encompassed within those of our experience, broadly understood.

Just this explains why it is that the range of philosophical agreement no greater. Since all rational philosophizing has to fall back on a variable manifold of probative support. For the fact of it is that there are always different alternative ways of reducing an inconsistent manifold of theses to consistency, and that the rational choice of one or the other of them requires a comparative standpoint of cogency and acceptability. And here in the end we arrive at an evaluative matter on which the different bodies of experience of different people are bound to lead them to different views.

But does this variability not reduce philosophy to a relativism that unravels its claims to be a rational enterprise?

Here, as in many contexts of deliberation it is important to distinguish with using four distinctive positions:
- *Personal contextualism:* Something relating to what is objectively appropriate and proper for (all) people falling within a particular circumstance, condition or category (the rights and duties of a ship's captain);
- *Idiosyncratic individualism:* something that holds only for a particular individual on the basis of circumstances pertaining uniquely to him/her alone (the rights and duties of the Prime Minister of Great Britain);
- *Indifferentist relativism:* something that roots in the specific wants, desires, wishes, preferences, and inclinations of particular individuals (the wine preferences of Thomas Jefferson);

- *Generalized universalism:* something that holds for anyone and everyone (the objectives of the Ten Commandments).

This situation exemplifies something general. Thus one's medical or dietary needs are matters of personalistic contextualism: what they hold for people in one's situation (but not for everyone). And yet are not matters of unfettered individual choice or taste. The issue is one of preferability rather than mere preference as such. The issues are matters of right/wrong, proper/improper, appropriate/inappropriate, and not matters of indifferent choice and personal taste.

There are various significant contexts of deliberations in which the important distinction between *personalistic contextualism* and *indifferent relativism* must be heeded with care. Oliver Cromwell maintained that what matters for people's best interests is "not what they want but what's good for them," And just this is at issue here.

And the crucially important consideration here is that the differential personalism of contextual variation is something that can be objective and rationally cogent. It need not be a rationally indifferent and arbitrary matter of personal taste subjective inclination and arbitrary preference.

If, for example, we are to have as an acceptable generalization that:
- Taking medicament is appropriate for anyone in the circumstance of condition C

then there has to be in the background a rationale of the type and format:
1. It is appropriate for anyone to take a medicament needed for (or optimally suited to) manage their medical condition;
2. Medicament M is optimally suited to manage the issues of people in circumstances C.

Therefore: Taking Medicament M is appropriate for anyone in the circumstances of condition C.

In the absence of such a way of validating a personalistically contextual position within the scope of a universal principle there is no way of establishing its legitimacy/appropriateness. In these contexts limited and contextualized appropriateness must inhere in and be demonstrative from an unlimited, generalized counterpart that can be validated on cogent principles of general acceptability.

Objectivity

And just here there emerges the important point that contextual appropriateness is something very different from mere taste and idiosyncratic preference. The personal need not be subjective: that there is room for an objective personalism of person-correlative fact that holds independently of the wants, wishes, preferences, tastes, and inclinations of individuals.

Thus you and I can differ in our needs—be they for medicaments, nourishments, information, etc. But these are facts about our objective condition make-up, and constitution that have nothing to do with our idiosyncratic wants, preferences, or inclinations. We differ in these matters, but those differences are objectively and factually based, correlative with aspects of our condition that are not subjectively idiosyncratic but actually realistic and as such accessible to all alike.

Contextualization is something very different from relativism. Granted, both alike make room for person-to-person causation. But unlike contextualism mere relativism is non-cognitive and psychological, it is a form of subjectivism, pulling rational cogency aside and leaving issue-resolution at the tender mercy of the idiosyncratic wants, wishes, decisions, performances, interactions of the individual. By contrast contextualization is objective, based on a cognitively cogent manners in the individual's objective considerations and circumstances—a matter not of wishes and desired and psychic interaction but of needs and requirements based on the objective fact of individual's situation. Relativeness is based on wants and idealisms; contextualities on circumstances and conditions.

And exactly this is at issue with philosophical personalism and provides for a an objectively based on the factual condition and circumstances of people, rather than reflecting their idiosyncratic wishes, wants, and preferences—exactly as one's biomedical condition, albeit-personal, is not arbitrarily and subjectively based on your wants, wishes, opinions, or preferences. In these regards there is indeed differences between persons, but in a way objectively based and rationalizable rather than subjectively indifferent as a mere matters of personal inclination, and preferences etc. Such a position is indeed variable between persons. But it is not arbitrary and mere subjective, but rationally founded in the objectively determinate circumstances and conditions of the individual. It is in sum a contextualized personalism that is objective and rational, and not a matter of an arbitrary, idiosymantically, feckless, and subjectivist indifferentism.

A common objection to pragmatism has it that the pragmatic theory of truth-acceptability calls for acceptance of a claim is justified when:
1. Action based on this claim is (generally) successful.

But—so the objection goes—this contention is never in fact cognitively avail-

able to us as such seeing that all that is ever actually at our cognitive disposition is:
2. Action based on this claim *is believed to be* (generally) successful.

And once this needed qualification is acknowledged, acceptance-justification proceeds by validating some of our beliefs by means of others, which converts pragmatism into a coherentism of sorts.[93]

The crux of the objection is that pragmatism is not in a position to address occurrences as such but is effectively confined to dealing with our *beliefs* about this. For in the end, the challenging "Tell us about what actually *is* the case rather than telling us about what you *believe* to be so" presents a challenge that really cannot be met. We simply have no cognitive access to the actual facts as such, in counter-distinction to our beliefs about them—access not mediated by what *is* so as contrasted with what *one thinks* to be so. And this inescapable circumstance constrains us from (1)-made ambition to (2)-mode qualification—a position that always substantiates beliefs in terms of others and so reverts to a mode of coherentism, something that pragmatism is designed to overcome.

Here, then, we have what might be called the Belief-Mediation Objection to a pragmatic theory of truth, resting on the pretty much unavoidable fact that what we have in hand is subjective when what we require is a matter of objectivity.

How can pragmatism possibly meet this objection?

From the very onset it has to be acknowledged that there is a decisive difference between making an objective claim of the format: "*p* obtains" and its subjective counterpart: "I believe/accept that *p* obtains." And it must be granted that if a pragmatic approach to truth-determination is to succeed, this will require a transit from premiss/inputs of the style of the former claim to conclusion/outputs of the style of the latter.

The crucial step towards a resolution lies in addressing the problem of epistemic gap between and fact via the crucial conception of *warrant*, accordingly insisting that what is at our cognitive disposal is not just (B) above but (C) Action based on this claim is *warrantly believed* to be (generally) successful.

To be sure, at this point the issue of warrant is invoked to bridge the epistemic gap between belief and fact, between subjectivity and objectivity, between appearance and reality. How can this work?

---

[93] The objection has been made in reviews of my *Methodological Pragmatism* (Oxford: Blackwell, 1977). And see also Amico 1993.

There are certain sorts of belief that are *virtually unmistakable*, belief where that crucial transit from acceptance to actual fact, and so from subjectivity to objectivity—is in order and appropriate. This crucial category of beliefs includes:
- *Subjective claims* ("I have a head-ache," "I believe that X," "I feel ill");
- *Reflexively ostensive claims* ("I exist," "I think," "I am trying to communicate," "This finger is wagging");
- *Blatantly obvious claims* ("That building is on fire," "That bomb is exploding," "that ball is bouncing," "The bridge has collapsed,: "Henry is walking");

With such virtually unmistakable claim on hand, pragmatic validation can overcome the objection that the successful implementation of belief is itself always merely a matter of belief, thus blocking pragmatism of being confined to a domain of subjective belief that falls short of factuality. Certain beliefs are not "mere" but inherently veritable. And it is these that will have to carry the burden of pragmatic work.

On this basis, we have a course of reasoning that shows that the statement issues for a pragmatic method that generates a class the vast majority of whose statements are "virtually unmistakable." This of course fits it out with an essentially intuitive rationale (albeit one based on pragmatic considerations). And this intuitive proceeding is itself backed not (or not only) by constructive proceeding but by fundamentally pragmatic considerations of a this-or-noting (or at least this-or-nothing-better) considerations. (Like any ultimate mode of validation, pragmatism must, in the end, prove to be self-relevant and self-substantiating.)

Pragmatism, in sum, can plausibly lay claim to effective implementation on issues whose resolution is otherwise impracticable.

Here one point must be noted with care. There is no logical necessity—no logical guarantee—that those claims here characterized as "virtually unmistakable" indeed are *always and invariably* correct. Their correctness is not an established fact but a plausible conjecture. For in this context, truth is a matter of rational presumption justified by situational need and not a demonstrable certainty. The certainty at issue is not absolute but virtual; mistakes here are to be seen not as impossible but only as highly unlikely, and the conclusion's truth is not an absolute certainty but rather a secure presumption. In other words we have here the justification of a policy as required by considerations of practical goal-oriented efficacy (this or nothing discernably better).

The substantiating argumentation here is not strictly speaking inductive in the positive format mode of "successful guidance of practice via applicative success. That is, we do not reason from success in earlier cases, to success in the present case as well. Rather, what we have here is the weaker "this or nothing better" mode of practical augmentation "there is nothing superior and more promising in sight than this: to all visible appearances its augers are at least as positive as those of any visible alternative." The branch of supportive argumentation thus itself belongs to the sphere of practical (rather than thematic) reasoning.

That crucial step of from subjectivity to objectivity a matter of constructive presupposition, a step of practical policy whose credentials are authorized as justified as evolution's gift to a creature that has to make its way over time in a challenging world through action guided by belief. What is at work here is a matter of pragmatic rather than logical necessitation—a pragmatically practical proceeding based on a rationale that is itself based on a homogeneously pragmatic rationale of validation.

Discord

Rational deliberations in philosophy is ultimately an exercise in question-resolution. Here is how the process does (or properly should) run:
1. One formulates the question to be addressed as definitively and clearly as possible;
2. One surveys the manifold of possible answers to the question at issue;
3. One enumerates and evaluates the comparative pros and cons—the assets and likelihoods of the various answers;
4. One seeks out that alternative which affords the most favorable outcome to the sort-effectiveness balance of the previous step;
5. One does what one can to resolve the question of the best available balance of pro- vs. con-consideration is sufficiently favorable to justify accepting this resolution as acceptable (or at least provisionally acceptable until such time as changed circumstance invite a reassessment of the matter).

Absent a resolution of the matter by a process of this structure, the best one can do is to suspend judgement on this matter until further notice.

At the center of this process is the combination of steps 3–4, the cost-benefit evaluations of the pros and cons of the alternative modes of issue resolution.

In the end, the pervasiveness of philosophical discord and disagreement does not contradict the objective cogency of the enterprise. It merely shows that the doctrinal position of people are (and properly should be) actually reflective of the body of their accessible experiences, experience where the experience of different people will invariably differ because the logic of the situation demands. Different premisses enjoin different conclusions.

14 The Pragmatic Perspective

The Problem of Pragmatism's Self-Validation

By its very nature, pragmatism stands committed to the idea that usefulness, efficacy in operation, application, and employment—is the hallmark of adequacy for anything that has a purpose to it. "By their fruits shall ye test them" is its motto. Pragmatism must accordingly meet the challenge "If as you teach, practical utility is the touchstone of merit, then what price your philosophical pragmatism itself—let alone the whole enterprise of philosophizing?"

So, if practical utility is to be our standard of validity, then where does this leave pragmatism itself? Indeed where does it leave all such matter of abstract theoretical reflection? If practical utility and applicative efficacy are to be our standard, then what price philosophy? What possible practical use—what pay off, to put it crudely, can possible some from endorsing positions on philosophical issues? If, as it so proudly claims, philosophy addresses "the big questions" of truth, justice, and beauty—if value and the ultimate nature of things—then does it not by its very nature cut itself off from life's practicalities to dwell in some lofty but ethereal region of its own? What "practical payoff" can possibly come from taking a position in such abstract matters so distinctly removed from the humdrum practicality of everyday life?

And this poses the question: how can pragmatic philosophy possibly be maintained on its own principles? After all, is not its own standard that of usefulness and practical efficacy, and is not philosophy the very quintessence of the impractical? Can pragmatism manage to deal effectively with the critic who complains: "Philosophy bakes no bread. The sort of abstract theorizing at issue here serves no *practical* purpose. Hence the entire enterprise is useless, pointless, illegitimate."

Knowledge as a Human Need

This line of thought reflects a very wrong-headed perspective. For philosophizing is part and parcel of the pursuit of knowledge, and this is a thoroughly purpose venture. After all, the sense of having a satisfactory comprehension of the way of things is essential for homo sapiens. In its total absence life could not be possible for us, and in its substantial absence could not find life satisfying. For us as thinking beings achieving some degree of comprehension is indispensable for realizing a manageable degree of physical and mental well-being. What is at issue here is something very practical that cuts to the core of our being as the kinds of creature that we humans indeed are.

We are a creature that can achieve neither physical nor mental well-being without feeling that we "know our way about" in the world to some significant extent. Cognitive accommodation to our surroundings is among our needs—and the demand for understanding is no less critical for our well-being than the demand for food. Feeling that we have an adequate grasp on the world about us is an unabandonable human need.

If information acquisition were not survival conducive we rational animals would not be here as the sorts of creatures we are, and could not long continue in existence as such. Here on earth, at least, intelligence is our peculiarly human instrumentality, a matter of our particular evolutionary heritage. Homo sapiens is also Homo quaerens. With us, the imperative to understanding is something altogether basic; we cannot function, let alone thrive, without information regarding what goes on about us. The knowledge that orients our activities in this world is itself the most practical of things—a rational animal cannot feel at ease in situations of which it can make no cognitive sense. The demand for understanding, for cognitive accommodation to one's environment, for "knowing one's way about," is one of the most fundamental requirements of the human condition. The discomfort of unknowing that drives us to inquire and investigate is a sentiment natural to any intelligent being and is readily understandable as such. We rational animals must feed our minds even as we must feed our bodies. In pursuing information, as in pursuing food, we have to make do with the best we can get at the time. We have questions and need answers—the best answers we can get here and now, regardless of their imperfections. This basic practical impetus to acquire coherent information represents the fundamental imperative of cognitive intelligence. Bafflement and ignorance (to give suspensions of judgment the somewhat harsher name that is their due) themselves exact a substantial price from us. The need for information, for cognitive orientation in our environment, is as pressing a human need as that for food itself—and more insatiable. We humans want and need our cognitive commitments to constitute an intelligible story, to give a comprehensive and coherent account of things. Cognitive vacuity or dissonance is as distressing to us as physical pain. Intelligence's pursuit of information—of putative knowledge- is one more facet of evolution's strategy of making what is useful to the species compelling to the individual by way of its own pleasure (want) or demand (need).

At the basis of the cognitive enterprise lies the fact of human curiosity rooted in the need-to-know of a weak and vulnerable creature emplaced in a difficult and often hostile environment in which it must make its evolutionary way by its wits. For we must act—our very survival depends upon it—and a rational animal must align its actions with its beliefs. We have a very real and material stake in securing viable answers to our questions as to how things stand in the world we live in.

The discomfort of unknowing is a natural human predicament. To be ignorant of what goes on about one is unpleasant to the individual and dangerous to the species from an evolutionary point of view. As William James wisely observed:

> The utility of this emotional affect of expectation is perfectly obvious; "natural selection," in fact, was bound to bring it about sooner or later. It is of the utmost practical importance to an animal that he should have prevision of the qualities of the objects that surround him.[94]

There is good reason why we humans pursue knowledge—it is our evolutionary destiny. Humans have evolved within nature to fill the ecological niche of an intelligent being. We are neither numerous and prolific (like the ant and the termite), nor tough and aggressive (like the shark). Weak and vulnerable creatures, we are constrained to make our evolutionary way in the world by the use of brainpower. It is by knowledge and not by hard shells or sharp claws or keen teeth that we have carved out our niche in evolution's scheme of things. The demand for understanding, for a cognitive accommodation to one's environment, for "knowing one's way about," is one of the most fundamental requirements of the human condition. Our *questions* form a big part of our life's agenda, providing the impetus that gives rise to our knowledge—or putative knowledge—of the world. Our species is *Homo quaerens*. We have questions and want (nay, *need*) answers.

In situations of cognitive frustration and bafflement we cannot function effectively as the sort of creature nature has compelled us to become. Confusion and ignorance—even in such "remote" and "abstruse" matters as those with which philosophy deals—yield psychic dismay and discomfort. The old saying is perfectly true: philosophy bakes no bread. But it is also no less true that man does not live by bread alone. The physical side of our nature that impels us to eat, drink, and be merry is just one of its sides. *Homo sapiens* require nourishment for the mind as urgently as nourishment for the body. We seek knowledge not only because we wish, but because we must. For us humans, the need for information, for knowledge to nourish the mind, is every bit as critical as the need for food to nourish the body. Cognitive vacuity or dissonance is as distressing to us as hunger or pain. We want and need our cognitive commit-

---

94 James 1897, pp. 78–9.

ments to comprise an intelligible story, to give a comprehensive and coherent account of things. Bafflement and ignorance—to give suspensions of judgment the somewhat harsher name they deserve—exact a substantial price from us.

The quest for cognitive orientation in a difficult world represents a deeply practical requisite for us. That basic demand for information and understanding presses in upon us and we must do (and are pragmatically justified in doing) what is needed for its satisfaction. For us, cognition is the most practical of matters. Knowledge itself fulfills an acute practical need. And this is where philosophy comes in, in its attempt to grapple with our basic cognitive concerns.

And then here is the larger picture of theoretical rather than simply basic knowledge. There are both individual and collective needs—recourses needed by individuals (food, shelter) and resources needed by social groups (laws, rules, social structures). The need for theorizing is of that latter sort. For a healthy and thriving society there has to be provision, there must be generic provision for it. It is a holistic heed that must be satisfied by some on behalf of all.

Philosophy as a Purposive Instrument

The history of philosophy consists in an ongoing intellectual struggle to develop ideas that render comprehensible the seemingly endless diversity and complexity that surrounds us on all sides. The instruments of philosophizing are the ideational resources of concepts and theories and it deploys them in a quest for understanding, in the endeavor to create an edifice of thought able to provide us with an intellectual home that affords a habitable thought shelter in a complicated and challenging world. As a venture in providing rationally cogent answers to our questions about large-scale issues regarding belief, evaluation, and action, philosophy is a sector of the cognitive enterprise at large.

And experience-based conjecture—prudent theorizing if you will—is the most promising available instrument for question-resolution in the face of imperfect information. It is a tool for use by finite intelligences, providing them not with the best *possible* answer (in some rarified sense of this term), but with the best *available* answer, the putative best that one can manage to secure in the actually existing conditions in which we do and must conduct our epistemic labors.

Despite those guarding qualifications, the "best available" answer at issue here is intended in a rather strong sense. We want not just an "answer" of some sort but a viable and acceptable answer—one to whose tenability we are willing to commit ourselves. The rational conjecture at issue is not to be a matter of *mere guesswork,* but one of *responsible estimation* in a strict sense of the term.

It is not *just* an estimate of the true answer that we want, but an estimate that is sensible and defensible: *tenable*, in short.

The need for such a truth-estimative approach in philosophy is easy to see. After all, we humans live in a world not of our making where we have to do the best we can with the limited means at our disposal. We must recognize that there is no prospect of assessing the truth—or presumptive truth—of claims (be they philosophical or scientific) independently of the use of our imperfect mechanisms of inquiry and systematization. And here it is *estimation* that affords the best means for doing the job. We are not—and presumably will never be—in a position to stake a totally secure claim to the definitive truth regarding those great issues of philosophical interest. But we certainly can—and indeed must—do the best we can to achieve a reasonable *estimate* of the truth. And *systematization* in the context of the available background information is nothing other than the process for making out this rationally best case. It is thus rational conjecture based on systematic considerations that is the key method of philosophical inquiry, affording our best hope for obtaining promising answers to the questions that confront us. C. S. Peirce observed this network aspect of philosophical systematization when he wrote as follows:

> Philosophy ought to imitate the successful sciences in its methods, so far as to proceed only from tangible premises which can be subjected to careful scrutiny, and to trust rather to the multitude and variety of its arguments than to the conclusiveness of any one. Its reasoning should not form a chain which is no stronger than its weakest link, but a cable whose fibers may be ever so slender, provided they are sufficiently numerous and intimately connected.[3]

The object of such an epistemic program is to use consideration of coherence as a guide to determine how smoothly and harmoniously a thesis can be enmeshed in the overall fabric of diverse and potentially discordant and competing contentions. We seek for the best candidates among competing alternatives—for that resolution for which, on balance, the strongest overall case can be made out. Accordingly, it is here not "the uniquely correct answer" but "the least problematic, most defensible position" that we seek. And the crux of our standard of acceptance lies, as we have seen, not with the issue of secure premises but with the issue of sensible conclusions—results that fit most smoothly and harmoniously within our overall commitment to the manifold "data" at stake in philosophical matters.

One further important point should also be stressed in this connection. To someone accustomed to thinking in terms of a sharp contrast between organizing the information already in hand and an active inquiry aimed at extending it, the idea of a *systematization of conjecture with experience* may sound like a very conservative process. This impression would be quite incorrect. Inquiry—and philosophical inquiry above all—must not be construed to slight the dynamical and ampliative aspect. And systematization itself is an instrument of inquiry—a tool for aligning question-resolving conjecture with the (of itself inadequate) data at hand. The factors of completeness, comprehensiveness, inclusiveness, unity, etc. are all crucial aspects of system, and the ampler the information-base, the ampler is the prospect for our systematization to attain them. The drive to system embodies an imperative to broaden the range of our experience, to extend and expand the database from which our theoretical triangulations proceed. In the course of this process, it may well eventuate that our existing systematizations—however adequate they may seem at the time—are untenable and must be overthrown in the pursuit of philosophy's goal of devising an amply comprehensively systematic framework of answers to our questions. Inflexible conservatism is not enjoined upon us here.

Philosophy has no distinctive information sources of its own. It has its own *problems*, but the *substantive materials* by whose means it develops answers must come from elsewhere. It thus has no distinctive subject-matter and furnishes no novel facts but only offers insights into relationships. For *everything* is relevant to its concerns, its tasks being to provide a sort of *expositio mundi*, a traveler's guidebook to reality at large. The mission of philosophy is to ask, and to answer in a rational and disciplined way, all those great questions about life in this world that people wonder about in their reflective moments.

What characterizes philosophy is neither a special subject matter nor a special methodology but rather—to reemphasize—its defining mission is that of co-ordinating the otherwise available information in the light of "big questions" regarding man, the world, and his place within its scheme of things. Philosophy deals largely with *how* and *whether* and *why* questions: how the world's arrangements stand in relation to us, whether things are as they seem, and why things should be as they are (for example, why it is that we should do "the ethically right" things). Ever since Socrates pestered his fellow Athenians with puzzling issues about "obvious" facts regarding truth and justice, philosophers have probed for the reason why behind the reason why.

After all it is, clearly, not just answers that we want, but answers whose tenability can plausibly be established; rationally defeasible and well-substantiated answers. And in particular this requires that we transact our question-resolving business in a way that is harmonious with and does not damage to—our prephilosophical connections in matters of everyday life affairs and of scientific inquiry. Philosophy's mandate is to answer questions in a manner that achieves overall rational coherence so that the answer we give to some of our questions square with those that we give to others. After all, philosophy is a venture in rational inquiry, a cognitive enterprise, a venture in question-resolution subject to the usual standards of rationality. In doing philosophy we are committed by the very nature of the project at hand to maintaining a commitment to the usual groundrules of cognitive and practical rationality.[95]

To be sure, we are sometimes said to be living in a post-philosophical age—an era when the practice of philosophy is no longer viable. But this is absurd. Nowadays more than ever we both desire and require the guidance of rigorous thinking about the nature of the world and our place within it. And the provision of such an intellectual orientation (*Orientierungswissen* as the German has it) is philosophy's defining mission. The fact is that the impetus to philosophy lies in our very nature as rational inquirers: as beings who have questions, demand answers, and want these answers to be cogent ones. Cognitive problems arise when matters fail to meet our expectations, and the expectation of rational order is the most fundamental of them all. The fact is simply that we *must* philosophize; it is a situational imperative for a rational creature.

But Why Risk It?

Yet why pursue such a venture in the face of the all too evident possibility of error? Why run such cognitive risks? For it is only too clear that there *are* risks here. In philosophizing, there is a gap between the individual indications at our disposal and the answers to our questions that we decide to accept. (As there also is in science—but in philosophy the gap is far wider because the questions are of a different scale.) Because of this, the positions we take have to be held tentatively, subject to expectation of an (almost certain) need for amendment, qualification, improvement, and modification. Philosophizing in the classical manner—exploiting the available indications of experience to answer those big questions on the agenda of traditional philosophy—is predicated on the use of reason to do the best we can to align our cognitive commitments with the substance of our experience. In this sense, philosophizing involves an act of faith: when we draw on our experience to answer our questions we have to proceed in the tentative hope that the best we can do is good enough, at any rate for our immediate purposes.

---

[95] There are, of course, very different ways of *doing* philosophy even as there are different ways of cooking food. But the enterprise itself is characterized by its defining objective: if one isn't doing that sort of thing, then one isn't pursuing it. (Sewing is not cooking food, nor is journalism philosophy.)

The question of intellectual seriousness is pivotal here. Do we care? Do we *really want* answers to our questions? And are we sufficiently committed to this goal to be willing to take risks for the sake of its achievement—risks of potential error, of certain disagreement, and of possible philistine incomprehension? For these risks are unavoidable—an ineliminable part of the philosophical venture. If we lose the sense of legitimacy and become too fainthearted to run such risks, we must pay the price of abandoning the inquiry.

This of course can be done. But to abandon the quest for answers in a *reasoned* way is impossible. For in the final analysis there is no alternative to philosophizing as long as we remain in the province of reason. We adopt some controversial position or other, no matter which way we turn—no matter how elaborately we try to avoid philosophical controversy, it will come back to find us. The salient point was already well put by Aristotle: "[Even if we join those who believe that philosophizing is not possible] in this case too we are obliged to inquire how it is possible for there to be no Philosophy; and then, in inquiring, we philosophize, for rational inquiry is the essence of Philosophy."[96] To those who are prepared simply to abandon philosophy, to withdraw from the whole project of trying to make sense of things, we can have nothing to say. (How can one reason with those who deny the pointfulness and propriety of reasoning?) But with those who *argue* for its abandonment we can do something— once we have enrolled them in the community as fellow theorists with a position of their own. F. H. Bradley hit the nail on the head: "The man who is ready to prove that metaphysical knowledge is impossible ... is a brother metaphysician with a rival theory of first principles."[97] One can abandon philosophy, but one cannot *advocate* its abandonment through rational argumentation without philosophizing.

The Pragmatic Dimension

In a way, pragmatism's approach to metaphysics is a position of intermediation. For the Realist, our presumed knowledge of the world grasps what is so "as such," in itself and independently of our cognitive proceeding, so that our knowledge is accordingly a matter of "correspondence with the real" (*adaequatio ad rem*) For the Idealist, our knowledge is a human construct, devised by us in our terms and for our purposes, independently of any reference to a thought-in-

---

96 Aristotle 1955, p. vii; for the text see p. 28. But see also Chroust 1969, pp. 48–50.
97 Bradley 1897, p. 1.

dependent reality. Pragmatism contrasts with both of these. Our knowledge is indeed a construct of ours devised in our own terms (so a point to the Idealists). But it is not a *free* construct because thought-independent reality "has the last word," so to speak, because how things work out—whether those man-designated menus kill or nourish, for example—is up to thought-independent realty and not ourselves (so a point to the Realists). And so as Pragmatists see it, our "reality" is a matter of a reality-picture produced in our own terms of reference but under the contrast of a mind-independent realty. Our intellection indeed proposes, but intellection-independent realty disposes and has "the last word."

A pragmatic approach thus enjoys significant advantage in its reliance on the objective realities of the human situation. Its recourse to effective practice opens the door to objectivity without absolutism, thus combining the most promising features of the traditional absolutism and present-day relativism in respect of knowledge. And the crucial fact is that such a meditative position allows a safe passage between the Scylla and Charybdis of problematic extremes.

The question "Should we philosophize?" accordingly receives a straightforward answer. The impetus to philosophy lies in our very nature as rational inquirers: as beings who have questions, demand answers, and want these answers to be as cogent as the circumstances allow. Cognitive problems arise when matters fail to meet our expectations, and the expectation of rational order is the most fundamental of them all. The fact is simply that we must philosophize; it is a situational imperative for a rational creature such as ourselves.

For Philosophy seeks to bring rational order, system, and intelligibility to the confusing diversity of our cognitive affairs. It strives for orderly arrangements in the cognitive sphere that will enable us to find our way about in the world in an effective and satisfying way. Philosophy is indeed a venture in theorizing, but one whose rationale is eminently practical. A rational animal that has to make its evolutionary way in the world by its wits has a deep-rooted need for speculative reason.

Needs can be divided into those required by and for individuals (food, shelter, clothing) and those required by and for entire communities (public order or road systems). And analogously there are those needs which can be met by individuals acting on their own (their dinner menu) or those which require public action for their satisfaction (personal security, say, or economic arrangements for credit and exchange), And the crux of the matter is that the development, possession, exchange and presentation of knowledge is concurrently requisite at all of these levels.

"But how can you speak of a human need for understanding given that many if not most people are altogether indifferent and oblivious to knowledge, inquiry, and theorizing—let alone matters of philosophy?" Part—but only part—of the answer here lies in the distinction between an individual and a communal need, between what people require individually (e.g. food and air) and what they need to have conjointly to function successfully as a collective community (e.g., law-and-order). But a no less important factor is that even as individuals we are better off—and better adjusted and able to live productively—when we have the conviction of understanding of how things work in the world.

Pragmatism as Self-Substantiating

Pragmatism takes the traditionalistic line of seeing the purposive character of the philosophical enterprise to lie in its very nature as a venture in seeking to answer our larger questions in a systematic way. But this purposive mission opens philosophizing up to the pragmatic standard of evaluation in terms of its efficacy and efficiency in realizing the purposes at issue. Because pragmatism sees question-resolving inquiry as a mode of praxis, it is able to take a fundamentally pragmatic view of the process of philosophy itself.

And this consideration brings into view one of pragmatism's important assets, namely its self-substantiation. For the sort of philosophizing at issue with pragmatism is of course itself a purposive enterprise—a venture in problem solving. And so it will be exactly to the extent to which the methodological pragmatism at issue here is able to provide a satisfactory mechanism for addressing the issues of epistemology, morality, etc., that is the crucial consideration for its own pragmatic validation.

Yet how does such a philosophical pragmatism itself fare by this pragmatic standard? The deliberations of the preceding discussion go to indicate, in some considerable detail, that it actually fares very well indeed.

Like any other rationally grounded human endeavor, philosophy too is a purposive endeavor, with an aim that sounds theoretical but is actually eminently practical, namely to orient our thinking with respect to those traditional "big questions" regarding ourselves, the world, and our place within its scheme of things. Since the days of its origin in classical antiquity, philosophers have seen the aim of the enterprise as one of providing answers to the "big questions" regarding mankind and its place in the world's scheme of things. So here, as elsewhere, adequacy will have to be maintained by success in purpose realization and the adequacy of a philosophy will accordingly have to be judged by its success in providing satisfactory answers to these "big questions." But how is this further factor of adequacy in issue resolution and question-answering being resolved?

Can the philosophical pragmatist live by his own teaching? Why on earth not? In practical and theoretical matters alike he can base his norms of appropriateness on considerations of functional efficacy. And clearly this criteriological schema holds for pragmatic philosophizing itself—it too can survive the scrutiny of the standard of satisfying the aims and purposed of philosophizing.

In the end, we seek to answer those "big questions" for a yet deeper ulterior purpose namely for the questions they provide for the conduct of life. "Philosophy is the guide of life" (*philosophia biou kybernêtês*) is the motto of this Phi-Beta Kappa society, and it is a well-intentioned precept. Those answers to our philosophical questions are—or should be—life-orienting: they provide our sense of direction and procedure in the management of life's affairs. And so the ultimate test of the adequacy of our philosophical positions and theories lies in the extent to which they guide life in a qualitatively satisfactory way. The best-available test of the adequacy of a philosophical position lies in the soundness of its furtherance to the quality of life of its exponents—the extent to which it contributes to their capacity to live a satisfying life.

Pragmatism has had a mixed reception in Europe. In Italy, Giovanni Papini and Giovanni Vailati espoused the doctrine and turned it into a party platform for Italian philosophers of science. In Britain, F. C. S. Schiller was an enthusiastic follower of William James, while F. P. Ramsey and A. J. Ayer endorsed pivotal aspects of Peirce's thought. Among Continental participants, Rudolf Carnap also put pragmatic ideas to work on issues of logic and philosophy of language and Hans Reichenbach reinforced Peirce's statistical and probabilistic approach to the methodology and proliferation of induction. However, the reception of pragmatism by other philosophers was by no means universally favorable. F. H. Bradley objected to the subordination of cognition to practice because of what he saw as the inherent incompleteness of all merely practical interests. G. E. Moore criticized William James' identification of true beliefs with useful ones—among other reasons because utility is changeable over time. Bertrand Russell objected that beliefs can be useful but yet plainly false. And various Continental philosophers have disapprovingly seen in pragmatism's concern for practical efficacy—"for success" and "paying off"—the expression of characteristically American social attitudes: crass materialism and naive populism. Pragmatism was thus looked down upon as a quintessentially American philosophy—a philosophical expression of the American go-getter spirit with its materialistic ideology of worldly success. But this is a poor caricature of philosophy and an inappropriate parody. For in matters of inquiry and theorizing "success" empathically does not lie in what is crass and materialistic domain, but in securing rationally acceptable answers to the questions that do—and rightly should—intrigue us as intelligent beings.

It should be seen as something close to self-evident that pragmatism is valid on its own terms. For the heart and core of this doctrine is that efficiency in meeting its objectives is the touchstone of adequacy with any purposive enterprise and that meeting our needs is a salient human enterprise. And in this light, with the need to satisfy our needs as a virtual tautology, a doctrinal stance to this effect is automatically in order.

To be sure, one key observation is crucial at this point, namely that our human needs also extend above and beyond the crass, material realm of physical well-being. For beyond the matter of *standard of living* there is also the matter of *quality of life*. Our human needs are not just material but cognitive, social, and even spiritual in nature. And in this light, pragmatism's stress on efficacy in need satisfaction means that the doctrine reaches out well above and beyond the crass and materials. A sensible pragmatism is not confined to the mundane basis of physical well-being and has no quarrel with the biblical doctrine that "Men doth not live by bread alone." To the contrary, a broader, more enlightened conception of the scope of human needs is crucial to any sensible form of pragmatism.

And so pragmatism is self-sustaining. In faring well by the standards that it itself endorses and enjoins, philosophical pragmatism enjoys the rational virtue of self-substantiation on this basis. It is a position willing and able to see its own status in the light of its own teachings, thereby practicing what it preaches in a way that skepticism, say, or relativism or scientism simply do not achieve. William James is quite right when he tells us that "I propose it [i.e., pragmatism] to my readers as something to be verified *ambulando*, or by the way its consequences may confirm it."[4] For pragmatism readily opens itself to be judged by the standard it applies elsewhere and that it can thus be seen as self-substantiating—as meeting that minimal requirement of a good philosophy of emerging as proper by its own standards.

It is, all too clearly, a decidedly positive feature of any philosophical theory that it be self-substantiating and sustaining—that it be tenable on its own telling by way of meeting the standards of cogency and appreciativeness that it itself advocates. The articulation of methodological pragmatism is designed to achieve this very end. This important fact that it is not only self-consistent but also ultimately proves to be self-supportive represents one of methodological pragmatism's most significant assets.

Functional adequacy is pragmatism's standard of merit: its encomium of approbation is "it works" and its criterion of merit is working out better than the available alternatives. This will of course leave open the question explanating question: why? What is it that explains functional/operational priority in the situation at hand? And here pragmatism's is emphatically not "Don't ask" with it prohibition not to look behind the veil of efficacy. It does not take the line that functional adequacy is an inaccessible sort of Kantian *Ding an sich ists* which one should not—cannot inquire. Pragmatism is entirely ready to contemplate an explanatory answer here. It is just that it sees that answer as something that can and should itself be developed by methods of inquiry whose validation is itself pragmatic. It thus sees pragmatism's own position as essentially self-sustaining, with the whole project validated in a way that is indeed cyclic not self-sustaining but not thereby as essentially circular. But the circle at issue is not viciously circular, but virtuously so. For it looks to the ultimately unavoidable consideration that any process of validating rational procedure must in the final analysis meet the demands of practical rationality.

But is this sort of self-validation not vitiated through circularity? Is it not useless for position or doctrine to be meritorious *on its own telling?* By no means.

In view of philosophy's commitment to systemic coherence, it is a crucial condition for adequacy that a philosophical position be self-sustaining—that its claims can be substantiated on its own theoretical telling. A philosophical position must practice what it preaches—unlike a radical, all-dismissive skepticism; for example, it must take a stance that is appropriate on its own telling. We would hardly be prepared to accept a philosophical position that was not able to qualify as well-substantiated even on its own principles. And as the deliberations of the present chapter indicate, pragmatism can stake a rightful claim to the possession of this indispensable merit. Since it fares well by its own standard, philosophical pragmatism enjoys the key rational virtue of self-substantiation. The sort of "circularity" at issue in self-substantiation is not something vicious and vitiating, but must, on the contrary, be adjudged altogether positive and beneficial. A position that is not appropriate on its own telling cannot sensibly be accepted as adequate.

Pragmatism's prioritization of matters of effective practice in the successful management of human affairs and the satisfaction of our needs and wants is thus noways at odds with the pursuit of theoretical reflection—including those relating to the range of philosophical issues that pragmatism itself addresses. For given the situation of man in the world's scheme of things this concern for knowledge—theoretical knowledge included—meets a critical human need. And so, on its own telling, pragmatism is accordingly self-substantiating in a way in which any adequate position in philosophy must ultimately be.

Bibliography

Addy, G. M., *The Enlightenment of the University of Salamanca* (Durham, NC: Duke University Press, 1966).
Amico, R. P., *The Problem of the Criterion* (Landham, MD: Rowman & Littlefield, 1993).
Aristotle, *Aristotelis Fragmenta Selecta*, ed. by W. D. Ross (Oxford: Clarendon Press, 1955).
Aristotle, *De Generatione Animalium*.
Blanshard, B., *The Nature of Thought* (London: Oxford University Press, 1939).
Bradley, F. H., *Appearance and Reality*. 2nd ed. (Oxford: Clarendon Press, 1897).
Bradley, F. H., *Essays on Truth and Reality* (Oxford: Clarendon Press, 1914).
Braun, L., *Historie de l'historie de la philosophie* (Paris: Editions Ophrys, 1973).
Brush, S. G., "Science and Culture in the Nineteenth Century," *The Graduate Journal,* Vol. 7, pp. 479–565, 1969.
Burrill, Donald R., *The Cosmological Arguments: A Spectrum of Opinion* (Garden City, NY: Anchor Books, 1967).
Campbell, J. K., "Hume's Refutation of the Cosmological Argument," *International Journal for the Philosophy of Religion*, Vol. 40, pp. 159–73, 1996.
Carus, P., *Kants Polegomena to any Future Metaphysics* (Chicago, IL: The Open Court Publishing, 1900).
Carnap, R., "Intellectual Autobiography," in P. A. Schilpp (ed.), *The Philosophy of Rudolf Carnap* (La Salle, IL: Open Court Publishing, 1963), pp. 54–55.
Casey, E. S., *Imagining: A Phenomenological Study*. 2nd ed. (Indianapolis, IN: University of Indiana Press, 2000).
Chisholm, R., *The Theory of Knowledge*. 2nd ed. (Englewood Cliffs, NJ: Prentice Hall, 1977), pp. 123–24.
Chroust, A.-H., *Aristotle, Protrepicus: A Reconstruction* (Notre Dame, IN: University of Notre Dame Press, 1969).
Clarke, S., *A Demonstration of the Being and Attribute of God* (London: Printed by Will. Botham, for James Knapton, 1705).
Craig, W. L., *The Cosmological Argument from Plato to Leibniz* (London: Macmillan, 1980).
Currie, G., *The Nature of Fiction* (Cambridge: Cambridge University Press, 1990).
Davidson, D., "On the Very Idea of a Conceptual Scheme," *Proceedings and Addresses of the American Philosophical Association*, Vol. 47 (1973), pp. 5–20.
Descartes, R., *Discourse on Method*.
Dilthey, W., *Gesammelte Schriften*, Vol. VIII (Stuttgart and Göttingen: Teubner and Vandenhoeck & Ruprecht, 1961).
Edwards, P., "The Cosmological Argument," *The Rationalist Annual for the Year 1959* (London: Pemberton, 1960).
Esposito, J. L., "Science and Conceptual Relativism," *Philosophical Studies*, Vol. 31 (1977), pp. 269–77.
Ewing, A. C., *Idealism: A Critical Survey* (London: Methuen, 1934).
Ewing, A. C., "The Correspondence Theory of Truth" *Non-Linguistic Philosophy* (London: Allen & Unwin, 1968).
Ferguson, N., *Virtual History* (New York: Basic Books, 1990).
Gale, R. M., *On the Nature and Existence of God* (Cambridge: Cambridge University Press, 1991).

Gallie, W. B., *Philosophy and Historical Understanding* (London: Chalk & Windus, 1964).
Gallie, W. B., "Essentially Contested Concepts," *Philosophy and Rhetoric*, Vol. 10 (1977), pp. 71–89.
Gonseth, F., "La Notion du normal," *Dialectica* Vol. 3 (1947), pp. 243–252.
Goodman, N., "The Way the World Is," *The Review of Metaphysics* (1960), pp. 48–56.
Goodman, N., *Fact, Fiction, and Forecast.* 2nd ed. (Indianapolis, IN: Bobbs-Merrill, 1965).
Haak, S., "[Review of] *The Strife Systems: An Essay on the Grounds and Implications of Philosophical Diversity* by Nicholas Rescher," *Philosophy and Phenomenological Research*, Vol. 48 (1987), pp. 167–70.
Hegel, G. W. F., *Vorlesungen über die Geschichte der Philosophie* (Berlin: Duncker und Humblot, 1836).
Hempel, C. G., "Science Unlimited," *The Annals of the Japan Association for Philosophy of Science* 14 (1973), pp. 187–202.
Horowitz, T., and G. Massey (eds.), *Thought Experiments in Science and Philosophy* (Savage, MD: Rowman & Littlefield, 1991).
Hume, D., *Dialogues Concerning Natural Religion* (London: 1779)
Hume, D., *A Treatise of Human Nature* (London: Dent, 1911).
Jackson, F. (ed.), *Conditionals* (Oxford: Clarendon Press, 1991).
Jacquette, D. (ed.), *Philosophical Essays: Classic and Contemporary Readings* (New York City, NY: McGraw-Hill, 2001).
James, W., *The Will to Believe and other Essays in Popular Philosophy* (New York City, NY: Longmans Green, 1897).
Joachim, H. H., *The Nature of Truth* (Oxford: Clarendon Press, 1906).
Johnstone, H. W. Jr., *Philosophy and Argument* (State College, PA: University of Pennsylvania Press, 1959).
Kant, I., *Critique of Pure Reason*.
Kekes, J., "Perennial Arguments," *Idealistic Studies*, Vol. 9 (1979).
Kekes, J., *The Nature of Philosophy* (Totowa: NJ: Rowman & Littlefield, 1980).
Kenny, A. (ed.), *Aquinas: A Collection of Critical Essays* (Notre Dame, IN: University of Notre Dame Press, 1976).
Kirk, G. S., and J. E. Raven, *The Presocratic Philosophers* (Cambridge: Cambridge University Press, 1957).
Leibniz, G. W., *Monadology*.
Levi, I., *For the Sake of Argument* (Cambridge: Cambridge University Press, 1996).
Lewis, C. I., *An Analysis of Knowledge and Valuation* (La Salle, IL: Open Court, 1962).
Lewis, D., *Counterfactuals* (Oxford: Blackwell, 1973).
Markovic, M., "Is Systematic Philosophy Possible Today?," in G. Ryle (ed.), *Contemporary Aspects of Philosophy* (Stocksfield/Boston, MA: Oriel Press, 1976), pp. 269–83.
McDermott, J. J. (ed.), *The Writings of William James* (Chicago, IL/London: University of Chicago Press, 1977).
Meinong, A., "Zur erkenntnistheoretischen Würdigung des Gedächtnisses" in A. Meinung, *Gesammelte Abhandlungen* (Berlin: Springer, 1933).
Moore, G. E., *Principia Ethica* (Cambridge: University of Cambridge Press, 1903).
Nagel, E., "Logic Without Ontology" in H. Feigl and W. Sellars (eds.), *Readings in Philosophical Analysis* (New York City, NY: Appleton-Century-Croft, 1949), pp. 191–210.
Newman, J. H., *An Essay in Aid of a Grammar of Assent* (London: Burns & Oates 1870).

Nietzsche, F., *The Geneology of Morals*, ed. by John McFarland (London: G. Allen, 1923).
Nute, D., *Topics in Conditional* Logic (Dordrecht/Boston, MA/London: D. Reidel, 1980).
Peirce, C. S., *Collected Papers*, ed. by C. Hartshorne and P. Weiss (Cambridge, MA: Harvard University Press, 1931–1934).
Plato, *Republic*.
Prado, C. G., *Making Believe: Philosophical Reflection on Fiction* (Westport, CT: Greenwood Press, 1984).
Pruss, A. R., "The Hume-Edwards Principle and the Cosmological Argument," *International Journal for Philosophy of Religion*, Vol. 434 (1988), pp. 149–65.
Ramsey, F. P., *The Foundations of Mathematics and Other Logical Essays*, ed. R. B. Braithwaite (London: Routledge, 1931).
Rashdall, H., *The Universities of Europe in the Middle Ages*, 2 vols. (Oxford: Clarendon Press, 1936; 2nd ed.).
Rescher, N., *Hypothetical Reasoning* (Amsterdam: North Holland, 1964).
Rescher, N., *The Coherence Theory of Truth* (Oxford: Clarendon Press, 1973).
Rescher, N., *The Primacy of Practice* (Oxford: Basil Blackwell, 1973b).
Rescher, N., *Plausible Reasoning* (Assen: Van Gorcum, 1976).
Rescher, N., "On Validating First Principles," *Allgemeine Zeitschrift für Philosophy*, Vol. 2 (1976b), pp. 1–16.
Rescher, N., *Dialectics* (Albany, NY: SUNY Press, 1977).
Rescher. N., *Methodological Pragmatism* (Oxford: Blackwell, 1977b).
Rescher, N., *Cognitive Systematization* (Oxford: Blackwell, 1978).
Rescher, N., "Philosophical Disagreement and Orientational Pluralism," *The Review of Metaphysics*, Vol. 32 (1978b), pp. 217–251.
Rescher, N., *The Riddle of Existence* (Lanham, MD: University Press of America, 1984).
Rescher, N., *The Strife of Systems* (Pittsburgh, PN: University of Pittsburgh Press, 1985).
Rescher, N., "Aporetic Method in Philosophy," *The Review of Metaphysics*, Vol. 41 (1987), pp. 253–297.
Rescher, N., "Thoughts Experimentation Pre-Socratic Philosophy," in T. Horowitz and G. J. Massey (eds.), *Thought Experiments in Science and Philosophy* (Savage, MD: Rowman & Littlefield, 1991), pp. 31–41.
Rescher, N., *Metaphilosophical Inquiries* (Princeton, NJ: Princeton University Press, 1994).
Rescher, N., *Philosophical Standardism* (Pittsburgh, PN: University of Pittsburgh Press, 1994b).
Rescher, N., "Metaphilosophical Coherentism," *Idealistic Studies*, Vol. 27 (1997), pp. 131–141.
Rescher, N., *Paradoxes* (Chicago/La Salle, IL: Open Court Publishing, 2000).
Rescher, N., "Philosophical Methodology," in Bo Mou (ed.) *Two Roads to Wisdom: Chinese and Analytical Philosophical Tradition* (Chicago, IL: Open Court Publishing, 2001).
Rescher, N., "What Is the Mission of Philosophy," in Dale Jacquette (ed.), *Philosophical Essays: Classic and Contemporary Readings* (New York City, NY: McGraw-Hill, 2001b).
Rescher, N., *Imagining Irreality* (Chicago/La Salle, IL: Open Court Publishing, 2003).
Rescher, N., "On Philosophical Systematization," *Southern Journal of Philosophy*, Vol. 43(2005), pp. 425–442.
Rescher, N., *Conditionals* (Cambridge, MA: MIT Press 2007).

Rescher, N., *What If?: Thought Experimentation in Philosophy* (Milton, UK: Taylor and Francis, 2018).
Rescher, N., and R. Brandom, *The Logic of Inconsistency* (Oxford: Basil Blackwell, 1979).
Roese, N. J., and J. M. Olsen, *What Might Have Been: The Social Psychology of Counterfactual Thinking* (Mahwah, NJ: Lawrence Erlbaum Associates, 1995).
Rorty, R., "Philosophy as a Kind of Writing: An Essay on Derrida," *New Literary History*, Vol. 10 (1978), pp. 141–160.
Rowe, W. L., "Two Criticism of the Cosmological Argument," in W. L. Rowe and W. Wainwright (eds.) *Philosophy of Religion: Selective Readings*. 2nd ed. (New York City, NY: Harcourt Brace Jovanavich, 1989), pp. 142–56.
Rowe, W. L., *The Cosmological Argument* (Princeton, N.J.: Princeton University Press 1975).
Russell, B., *Principia Mathematica* (Oxford: Oxford University Press, 1910).
Russell, B., *The Problems of Philosophy* (London: Oxford University Press, 1912).
Schiller, F. C. S., *Must Philosophers Disagree? and Other Essays in Popular Philosophy* (London: Macmillan, 1934).
Schlick, M., "The Turning Point in Philosophy," in A. J. Ayer (ed.), *Logical Positivism* (Glencoe, IL: Free Press, 1959), pp. 53–59.
Schlick, M., "The Foundation of Knowledge," in A. J. Ayer (ed.), *Logical Positivism* (Glencoe, IL: Free Press, 1959b), pp. 209–27.
Seneca, *Epistles*.
Sextus Empiricus, *Outlines of Pyrrhonism*, translated by R. G. Bury (Cambridge, Massachusetts: Harvard University Press, 1933).
Sidgwick, H., *Methods of Ethics* (London and New York: Macmillan, 1874).
Simon, H. A., "Thinking be Computers," in R. G. Colodny (ed.) *Mind and Cosmos* (Pittsburgh, PN: University of Pittsburgh Press, 1966).
Simon, H. A., "Scientific Discovery and the Psychology of Problem Solving," in R. G. Colodny (ed.) *Mind and Cosmos* (Pittsburgh, PN: University of Pittsburgh Press, 1966).
Simon, H. A., "The Architecture of Complexity," in H. A. Simon, *The Sciences of the Artificial* (Cambridge, MA: MIT Press, 1989), pp. 193–229.
Sorensen, R. A., *Thought Experiments* (Oxford: Oxford University Press, 1992).
Wallace, K., "Metaphysics and Validation" in T. Rockmore and B. J. Singer (eds.), *Antifoundationalism Old and New* (Philadelphia, PN: Temple University Press, 1992), pp. 209–238.
White, A. R., "Coherence Theory of Truth," in P. Edward (ed.), *The Encyclopedia of Philosophy*, Vol. 2 (1967).
Weitz, M., *The Opening Mind* (Chicago, IL: University of Chicago Press, 1978).
Wigner, E. P., "The Unreasonable Effectiveness of Mathematics in the Natural Sciences," *Communications on Pure and Applied Mathematics*, Vol. 13 (1960), pp. 1–14.
Ockham W. of, *Philosophical Writings*, ed. by P. Boehner (Edinburgh: Nelson, 1957).
Winch, P., *The Idea of a Social Science* (London: Routledge, 1958).
Winch, P., "Understanding a Primitive Society," *American Philosophical Quarterly*, Vol. 1 (1964), pp. 307–25.
Woods, J., *The Logic of Fiction* (The Hague: Mouton, 1974).
Woods, M., *Conditionals* (Oxford: Clarendon Press, 1997).

Index of Names

Addy, G. M., 153n62, 211
Amico, Robert P., 195n94, 211
Aquinas, St. Thomas, 1, 155
Aristotle, 1, 31n4, 32, 47, 66 93, 107, 138, 140, 146, 150–151, 171, 205, 205n97, 211
Aurelius, Marcus, 48, 90, 92
Ayer, A. J., 208, 214

Barth, Karl, 155
Berkeley, George, 181, 185
Blanshard, Brand, 71–72, 72n26, 211
Bradley, F. H., 43, 67, 72, 72n27, 72n28, 73, 118, 118n43, 119n44, 205, 205n98, 208, 211
Brandom, Robert, 38n8, 214
Braun, Lucian, 84n36, 211
Brush, Stephen G., 129n54, 211
Burrill, Donald R., 211

Campbell, Joseph K., 211
Carnap, Rudolf, 166n77, 173, 174n83, 208, 211
Carus, Paul, 211
Casey, Edward S., 211
Cassirer, Ernest, 1, 168n79
Cezanne, Paul, 164n74
Chisholm, Rudolf, 160n68, 211
Chroust, Anton-Hermann, 205n97, 211
Clarke, Samuel, 211
Collingwood, R. G., 156, 158
Colodny, R. G., 214
Craig, William Lane, 211
Cromwell, Oliver, 193
Currie, Gregory, 211

Davidson, Donald, 183n91, 211
Descartes, René, 53, 88, 112, 120, 140, 154, 156, 157n64, 158, 169, 211
Dewey, John, 1, 29, 46
Dilthey, Wilhelm, 93n37, 103, 104n39, 211
Durer, Albrecht, 164n74

Edwards, Paul, 211

Epictetus, 48, 90, 92
Epicurus, 102
Epimenides, 161
Esposito, Joseph L., 166n77, 211
Ewing, A. C., 72n28, 75n33, 211

Ferguson, Niall, 211

Gale, Richard M., 211
Galen, 82, 185
Gallie, W. B., 159n67, 212
Gassendi, José, 185
Gonseth, Ferdinand, 70n24, 212
Goodman, Nelson, 164n74, 167, 212
Green, T. H., 118

Haak, Susan, 65n19, 212
Hegel, G. W. F., 83–84, 84n36, 102, 104–105, 113, 118, 124, 132, 141, 154–155, 168n78, 170, 212
Hempel, Carl G., 212
Heraclitus, 89, 99–100, 124
Herbart, Johann Friedrich, 102–104
Hippias, 93
Holbein, Hans, 164n74
Horowitz, Tamara, 212–213
Hugo, Victor, 189
Hume, David, 81, 120, 120n45, 122, 122n49, 157, 157n65, 212
Huxley, T. H., 129n54

Jackson, Frank, 212
Jacquette, Dale, 1n, 212–213
James, William, 4, 35, 46, 164–165n76, 170–171, 172n81, 176, 200, 200n95, 208–209, 212
Joachim, H. H., 75–76n33, 212
Johnstone, Henry W. Jr., 66n20, 123n50, 161n69, 168n78, 212

Kant, Immanuel, 1, 7, 9, 81, 87, 118, 120–121, 141, 154–155, 157n66, 161n71, 190, 212
Kekes, John, 6n1, 159n67, 212

Kelvin, William Thomson, 129n54
Kenny, Anthony, 212
Kierkegaard, Søren Aabye, 124
Kirk, G. S., 124, 124n51, 212

Leibniz, Gottfried Wilhelm, 53, 118, 120, 141, 154, 212
Leslie, John, 140
Levi, Isaac, 212
Lewis, C. I., 73n29, 73n30, 212
Lewis, David, 212
Lucretius, 102

Mandeville, John, 189
Manet, Edouard, 164n74
Markovic, Michaili, 180n86, 212
Massey, Gerald J., 212–213
Mates, Benson, 161n71
McDermott, J. J., 165n76, 171n80, 176n84, 212
Meinong, A., 73n29, 212
Mill, John Stuart, 1,
Moore, G. E., 141, 141n59, 208, 212

Nagel, Ernest, 45, 45n12, 212
Nelson, Horatio, 77
Newman, John Henry, 155, 165n76, 212
Nietzsche, Friedrich Wilhelm 1, 121–122, 122n48, 124, 154–155, 213
Nute, Donald, 213

Ockham, William, 214
Olsen, James M., 214
Ortega y Gasset, Jose, 124

Pascal, Blaise, 30
Peirce Charles Sanders, 33, 33n5, 41, 41n10, 133, 157n66, 158, 202, 208, 213
Picasso, Pablo Ruiz, 164n74
Plato, 1, 21, 48, 59, 66, 88, 90, 92, 96–97, 112, 140, 164n72, 185, 213
Prado, C. G., 213
Pruss, Alexander R., 213
Pythagoras, 100, 124

Ramsey, Frank Plumpton, 59, 60n15, 98, 98n38, 161n69, 208, 213

Rashdall, Hastings, 153n62, 213
Raven, J. E., 124, 214n51, 212
Rescher, Nicholas, 15n3, 37n7, 38n8, 64n17, 68n21, 70n23, 70n24, 70n25, 120n46, 125n52, 131n55, 132n56, 164n75, 181n88, 183n92, 212–14
Roese, Neil J., 214
Rogers, Will, 106
Rorty, Richard, 214
Rowe, William L., 214
Russell, Bertrand, 74, 74n31, 75, 105, 105n41, 108, 161, 162–163n72, 208, 214

Schiller, F. C. S., 165n76, 208, 124
Schlick, Mortiz, 74n31, 157, 157n66, 124
Schopenhauer, Arthur, 124
Seneca, 186n93, 214
Sextus Empiricus, 48, 156n63, 173n82, 214
Shakespeare, William, 2
Sharukie, A., 164n74
Sidgwick, Henry, 141, 141n60, 214
Simon, Herbert A., 41n9, 70n24, 214
Socrates, 59, 66, 97, 203
Sorensen, Roy A., 214
Spinoza, Baruch, 53, 66n20, 95, 140, 154, 161

Thales, 14, 99–100
Thompson, William, 129n54

Unamuno, Miguel de, 124

Wallace, Kathleen, 64n18, 214
Weitz, Morris, 159n67, 214
White, A. R., 74n32, 214
Whitehead, Alfred North, 141
Wigner, Eugene P., 128–129, 128n53, 214
Winch, Peter, 166n77, 214
Wittgenstein, Ludwig, 123
Wolff, Christian, 105, 105n40
Woods, John, 214
Woods, Michael, 214

Xenophanes, 124

Zeno, 95, 96, 100

www.ingramcontent.com/pod-product-compliance
Lightning Source LLC
Chambersburg PA
CBHW031312150426
43191CB00005B/193